I0880276

Basque Immigrants and Nevada's Sheep Industry

The Basque Series

Basque Nationalism
by Stanley G. Payne

Amerikanuak: Basques in the New World
by William A. Douglass and Jon Bilbao

Aurrera!: A Textbook for Studying Basque
by Linda White

Back to Bizkaia: A Basque-American Memoir
by Vince J. Juaristi

The Basques, the Catalans, and Spain:
Alternative Routes to Basque Nationalization
by Daniele Conversi

Chorizos in an Iron Skillet: Memories and Recipes
from an American Basque Daughter
by Mary Ancho Davis

The Circle of Mountains: A Basque Shepherding Community
by Sandra Ott

Deep Blue Memory
by Monique Laxalt Urza

An Enduring Legacy: The Story of Basques in Idaho
by John Bieter and Mark Bieter

Gernika, 1937: The Market Day Massacre
by Xabier Irujo

The Good Oak
by Martin Etchart

Reclaiming Basque: Language, Nation, and Cultural Activism
by Jacqueline Urla

The Basque Language
by Alan R. King

Basque Firsts
by Vince J. Juaristi

Basque Immigrants and Nevada's Sheep Industry
by Iker Saitua

BASQUE IMMIGRANTS AND NEVADA'S SHEEP INDUSTRY

Geopolitics and the Making of an Agricultural Workforce, 1880–1954

IKER SAITUA

UNIVERSITY OF NEVADA PRESS *Reno & Las Vegas*

University of Nevada Press | Reno, Nevada 89557 USA
www.unpress.nevada.edu

Cover design by Rebecca Lown Design
Cover photographs courtesy of Iker Saitua

LIBRARY OF CONGRESS CATALOGING-IN-PUBLICATION DATA

Names: Saitua, Iker, 1987– author.
Title: Basque immigrants and Nevada's sheep industry : geopolitics and the making of an agricultural workforce, 1880–1954 / Iker Saitua.
Other titles: Basque series.
Description: Reno ; Las Vegas : University of Nevada Press, [2019] | Series: Basque series | Includes bibliographical references.
Identifiers: LCCN 2018041480 (print) | LCCN 2018050926 (ebook) | ISBN 9781943859993 (pbk. : alk. paper) | ISBN 9781948908016 (cloth : alk. paper) | ISBN 9781948908023 (ebook)
Subjects: LCSH: Sheep industry—Nevada—History. | Basque Americans—Nevada—History. | Shepherds—Nevada—History. | Sheepherding—Nevada—History. | Immigrants—Nevada—History. | Sheep industry—West (U.S.)—History.
Classification: LCC HD9436.U53 S25 2019 (print) | LCC HD9436.U53 (ebook) | DDC 331.6/24660979309041—dc23
LC record available at https://lccn.loc.gov/2018041480

The paper used in this book meets the requirements of American National Standard for Information Sciences—Permanence of Paper for Printed Library Materials, ANSI/NISO Z39.48-1992 (R2002).

FIRST PRINTING

Manufactured in the United States of America

This book is dedicated
to the memory of my late grandfathers,
Manuel Saitua-Torrontegi (1922-1993)
Imanol Idarraga-Monasterio (1930-2017),
my late grandmother,
Mari Carmen Brasa-Zenikazelaia (1924-2016),
and my late great-aunt,
Sole Brasa-Zenikazelaia (1928-2017).

The four of them were members of a generation
who survived the Spanish Civil War
and its consequences in the Basque Country.

Contents

Note for Readers

Translations from sources in Basque, Spanish, and French languages including titles of secondary sources are the writer's. Generally, the book presents Basque terms and names in anglicized forms, but Basque terms that have not been anglicized appear in their original spelling. The original Basque names are presented as they appear in the official documents, despite evident misspellings. It should be noted that typically Basque immigrants on their arrival in the United States anglicized their names and sometimes adopted a new one in their American integration process. Further, in the case of the Basque immigrant women, they lost their family name by marriage in the United States.

The present study considers the Basque Country as a geographical territory consisting of the following seven provinces which are divided between Spain and France. In Spain, the Basque provinces are: Biscay, Guipúzcoa, Álava, and Navarre. In France, the provinces are: Labourd, Lower Navarre, and Soule.

Introduction: The Basque Frontier of the American West

On June 20, 1937, the *Oregonian* published in its Sunday magazine a small photographic report about Basque immigrants in southeastern Oregon. The Basques were widely identified with the western open-range sheep industry, and the article explained how the Basque immigrant community was highly concentrated in both Malheur and Harney counties because their livelihoods depended on the sheep ranching economy in this corner of the Interior West. Moreover, in some places, according to the article, Basque immigrants outnumbered the native-born residents. For instance, the report estimated that in the little frontier town of Jordan Valley, Basque immigrants and their children represented about 66 percent of the total population.

The Basque were, according to this article, "a friendly, hard-working race." The article continued: "In northeastern Spain, from the Bay of Biscay back into the Pyrenees mountains, is the homeland of this unique race. Since prehistoric times this has been their native heath. An intensely proud and independent people, whose origin is cloaked in mystery, theirs is a race apart.[1]

The words constituted the typically positive view of Basque immigration in the United States as the narrative had developed during previous decades when race classifications abounded. The article praised the Basque immigrants for their alleged industriousness in sheep grazing, warmhearted hospitability, and sense of pride in their origins. It also asserted that Basque immigrants assimilated easily into the dominant American culture. Although Basques were not Anglo-Saxon immigrants, the article said, "they are masters of English after but few years in this country," going on to describe their physical characteristics in ways that were meant to assert affinities with other dominant social and racial groups, and thus indicating how much by the late 1930s the Basques were well on their way to becoming an accepted, if not welcomed, ethnic group in the American West.[2]

But this accepted social status was not always the case for the Basques in the West, and their journey as an immigrant group in America—particularly within the western open-range sheep industry that was at the heart of Basque immigration and economic involvement from the late nineteenth century to the second half of the twentieth century—was never a straightforward one. Indeed, the history of Basque immigrants working in the American sheep industry reveals a rich and complex story of socioeconomic integration within changing American and international political and social contexts.

▲ ▲ ▲

Basques in the American West were associated with the open-range sheep industry from the 1880s to the 1970s. Since its beginnings, in one way or another, Basque immigrants played a noted role in the development of the commercial sheep raising in the West as a primary labor force. In the 1890s, as sheep ranching expanded rapidly in the West, the Basque Country became an important source of pliable labor to work in the public rangelands. At the turn of the century, although other ethnic groups also entered this occupation, Basques had become a noticeable and visible group in sheep grazing, particularly in Nevada, which was a nexus for Basque immigrant work in this industry. Even though Basque immigrants also worked in other industries besides sheep grazing, their work in this industry in particular shaped and strengthened the broader Basque-American community—or "Amerikanuak" as they were called, a corruption of the word "Americans" that came to refer to "Basque-Americans"—all around the West.[3] This historical process, which is the main subject of this study, was a complex, multidimensional, and sometimes contradictory one.

After the prominent and continuous Basque presence in the current Southwest region of the United States during the Spanish colonization, Basque immigrants began appearing in North America in considerable force around the late nineteenth century.[4] Basque immigration to the United States is largely considered to have begun around the time of the gold rush, when a number of Basques living in Latin America migrated north to California.

By the late 1880s, though their numbers were relatively few, a handful of Basque families, such as the Altubes, who had initially arrived via Argentina, had created an immigrant enclave in Nevada based largely on the open-range sheep industry.[5] Taking advantage of the free and open range available in the state at that time, these Basque pioneers built up quickly thriving sheep operations, which opened a floodgate for further Basque immigration as the open-range sheep industry became an immigrant lure for the Basques. This resulted in an expanding process of chain migration, which channeled Basque newcomers into the bottom ranks of the sheep industry and promoted the development of kindred networks.[6] Thereof, an occupational concentration process occurred with the arrival of more Basque immigrants who settled nearby and worked as sheepherders along with their countrymen in the American West.[7]

Despite their growing numbers in the sheep industry in Nevada, Basque immigrants who arrived in the late nineteenth century were newcomers to an already established grazing empire, one that was based largely on cattle. In the 1890s, when an increasing number of Basque immigrants arrived in the West to work as sheepherders, their presence in the public ranges began to disturb the economic interests of the older cattle ranchers. Moreover, the 1890s witnessed a remarkable boom in the sheep business of the Great Basin, especially after the disastrous winter of 1889–1890, which wiped out huge percentages of livestock and vegetation. After such an extreme-weather event, which forced livestock operators to change their grazing practices (particularly secure provisions for winter feed), the economics of sheep grazing seemed to fare better than cattle. Consequently, the number of sheep outfits multiplied and rangelands became more fully stocked. Then, competition between cattle owners and sheep graziers for the forage resources intensified, oftentimes flaring into violence. Basque sheepherders' frequent trespassings on private land or crossings on public-domain lands occupied by others, intentionally or unintentionally, resulted oftentimes in tragic consequences for many Basque immigrant sheepherders.

The issues facing the livestock industry and the Basque sheepherders in the West were not merely state or local issues, and increasingly, the livestock industry and land use became a focus of the federal government. By the end of the nineteenth century, Congress began withdrawing public-domain lands—largely in the American West—from private entry and acquisition for the purpose of protecting the country's natural resources, primarily water and timber. Consequently, the question of resource use became an important one. Congress would go on to authorize the use of resources, including grazing on these lands, in ways that eventually brought regulations that directly affected the itinerant sheep grazing—in which the Basque were particularly involved—on Nevada's open ranges. During the late nineteenth and early twentieth centuries, just as Congress and the president moved to establish forest reserves and eventually national forests, issues of rangeland governance made the question of the Basque itinerant sheepherders central to a political debate at the state and national levels over who could use public grazing lands.

Although Basque immigrants in the West working in the sheepherding industry had generally been well considered by the dominant social and economic classes during the nineteenth century, older cattlemen—in the context of increased economic competition and social tension on Nevada's ranges—began to scapegoat Basque sheepworkers for all the problems with grazing on public-domain lands. Along with the general nativist sentiment in the United States during this era, the years around World War I would bring with them discrimination and efforts to exclude Basques from the public-domain lands.

Indeed, during the first decades of the twentieth-century, just as Basques began to become settled in the Interior West with some of them building prosperous sheep-dealing businesses, an anti-Basque movement was begun by cattle ranchers, conservationists, and some politicians. Basque immigrant sheepherders, with their flocks coming onto public grazing lands, were perceived as a serious threat to the cattle economic interests, sustainable agricultural development, and public interest in general. They were blamed for a number of interrelated economic and environmental problems affecting the public grazing lands. The increasing presence of Basque sheepherders in the public rangelands drew derision amidst an American cowboy culture in the West.

This crusade against the sheepherders would ultimately create the most pernicious image of Basque immigrants in the West—as the antithesis of the nationally venerated image of cowboys. While cowboys were viewed as the embodiment of the American *frontier* spirit of self-reliance by hardworking men who rode horses, sheepherders were denigrated as despicable and as laborers of an inferior class who walked long distances allowing sheep to roam at will across the wide-open range. The elevation of the venerable cowboy would seem, in some ways, to come at the expense and downgrading of the humble sheepherder. The social stigma that began to be carried by the sheepherding job would be enabled and compounded by racial prejudice and discrimination, something that the Basques could not escape.

These derogatory stereotypes of Basques would not only infiltrate disputes among livestock operators and communities in the West; they would also be used by politicians and industry lobbyists seeking advantages among state and federal land policy, specifically for cattle ranchers. Indeed, racialized perspectives of Basque sheepherders, and thus of Basque immigrants more generally, would characterize industry lobbying and Basque immigration politics for decades.[8] Nevertheless, it would not always be against the Basques. Even while being subjected to overt racial and ethnic discrimination among cattle ranchers, Basque immigrants would, at the same time, find themselves placed at the top of the racialized sheep agricultural labor hierarchy.

In the early twentieth century, there was a strong American public image of Basque immigrant workers as possessing certain racial/cultural advantages necessary for sheepherding that nobody else had. Basques had acquired a distinctive reputation for an expertise and pioneering skill in the sheep industry of Nevada. It was a perspective, which, along with the racialization of the Basques in the late nineteenth and early twentieth centuries, would greatly contribute to the acceleration of those immigrants' integration process.

Through the Anglo-American lens, Basques were socially constructed as an ancient, unique, and mysterious race from Europe. The vagueness of the racial construction surrounding Basques gave them distinctiveness which was reflected in their sheepherder identity. The increased ubiquity of Basque immigrants in the sheep industry resulted in the construction of a cultural stereotype: the Basque was a qualified, dependable, and good sheepworker. Basques' alleged racial difference set them off from some other ethnic groups involved in the sheep grazing occupation of the American West. Such constructed racial identity would be constantly reflected, reinforced, and reiterated in the various representations of Basque sheepherders in the West in newspapers, popular literature, and in the words of politicians and lobbyists throughout the twentieth century.

The largest influx of Basque immigrants from the 1890s to the 1920s coincided with what is known as the "new immigration" from southern and eastern Europe.[9] Like many other European immigrant communities, Basques were induced to immigrate to the American West in the context of the American industrial and agricultural expansion in the late nineteenth century. And like its non–English speaking European immigrant counterparts, Basque immigration also was disrupted after the passage of the Immigration Act of 1924.[10]

The restrictive legislation of the 1920s constrained Basque immigration to the American West. Particularly, the Immigration Act of 1924 established quotas as principal means of regulating immigration, which limited significantly immigration from southern and eastern European countries. Because the Basque Country is divided between Spain and France, the 1924 immigration law affected Basques differently depending on their country of origin. The annual immigration quota for Spain was much smaller than that of France. After 1924, then, Basques from Spain faced greater challenges and had fewer opportunities to enter the United States than those emigrating from France, something that eventually would have major consequences for the recruiting of Basque immigrant labor in the sheep industry.

During the interwar period, a major shortage of "skilled" sheepherders occurred due to the 1924 immigration law, which had greatly reduced opportunities for Basque immigration to the United States. As a result of this shortage, negative perceptions of Basque sheepherders that emerged previously when they were seen as threats to the cattle industry would give way to positive perceptions, and the issue of Basque immigration would become intertwined with important labor and political debates of the first half of the twentieth century. Nevada and western woolgrowers reproduced the stereotype of the *good sheepherder*; a conception they used to justify a dependable Basque immigrant labor force to tend their sheep and retain control over such labor. In that way, furthermore, they racialized the production relations in the industry to justify the necessity to bring more Basque immigrant workers to Nevada's sheep industry: their desirable immigrant labor.

With the outbreak of World War II, the rapid industrial development in Nevada and the West created new job opportunities on a large scale, which former Basque immigrants living in the United States found more attractive than tending sheep. Because of the longstanding belief among American sheep ranchers that only Basque workers could effectively handle their labor needs, a group of Nevada ranchers took their demands to import Basque immigrant workers to Congress. Senator Patrick McCarran of Nevada became a leading voice for allowing recruitment of increasing numbers of Basque immigrants from Spain. From 1942 until the time of his death in 1954, McCarran lobbied persistently to bring Basque immigrants to Nevada.

Basque immigration raised issues of policy implications for both the United States and Spain, which became particularly relevant during and after World War II. Although the disastrous consequences of the Spanish Civil War (1936–39) had left many Basques eager to leave Spain, the quota system in the United States reduced opportunities for emigration to the United States, despite the labor shortage in the sheepherding industry in the American Far West which had traditionally employed many Basque immigrants. Furthermore, the unfriendly relations between the United States and Francisco Franco's Spain during World War II and the early Cold War impeded any agreement that would have sponsored the importation of Basque immigrant laborers in the United States.

The recruiting process of Basque agricultural labor would ultimately unfold in the context of the foreign relations between the United States and Francisco Franco's dictatorship from the Second World War to the Pact of Madrid of 1953. Even though the United States government postponed the signing of an official agreement with Franco's regime until 1953, the ongoing lobbying by Nevada's sheep ranchers and Senator McCarran's influence in Congress enabled the issue of the recruitment of Basque immigrants to become integrated within broader informal bilateral trade and commercial negotiations including the recruitment of Basque immigrants. The Pact of Madrid normalized diplomatic relations with Franco's regime, which in turn served to strengthen other commercial relations including the recruitment of Basque immigrant laborers.

The major era of Basque involvement in sheepherding in the American West took place over a long period of time, from the 1870s to the 1950s. For much of that history, particularly in the early twentieth century, Basques were frequently perceived as having traditions as experienced sheepherders in their home country, an idea that was perpetuated and interestingly utilized for years. But this idea was not true. When Basque immigrants came to the American West to work in the sheep industry, they entered an unfamiliar occupation, in an unfamiliar geographical context, and adapted as they could. In the large, complex, and diverse geography of the American West, the varied locations and circumstances made Basque sheep workers vulnerable to many social, economic, legal, and environmental challenges, something which can be seen throughout the history of their involvement in the industry.

I have used the term "sagebrush laborers" to refer to the Basque immigrant sheepherders. This term recognizes Basque immigrant sheepherders as workers who produced an important commodity as part of the capitalist agricultural industry of Nevada, while rejecting the romanticized image of the good Basque sheepworker who—in contradistinction to most low-paid workers in agriculture—conformed so closely to an idealized vision of the good "white American." Sagebrush prevailed throughout much of the grazing grounds of Nevada's open rangelands where the Basque worked. Both Basque and sagebrush were ubiquitous in the sheepherding environment, and together they rendered to the Nevada ranges a particular identity. The Basque laborer and the open ranges of Nevada where this sagebrush grew made the sheep industry viable. The one provided cheap labor from a faraway land, and the ranges provided virtually free grazing. The labor was marginal as well as the land. These two parts of an economic equation of scale made possible the rise of Nevada's sheep industry from the 1890s onward.

▲ ▲ ▲

European immigrants have been largely ignored by the traditional and new western historiography. In the last years, however, scholars have begun a movement toward a more integrative study of European immigrants in western history.[11] If the new European immigration has been overlooked in the history of the West, this is especially true of the Basque immigrant community.[12]

In 1893, Frederick Jackson Turner presented his thesis "The Significance of the Frontier in American History" to the American Historical Association in Chicago. Turner's influential ideas contended that the frontier process produced democratic institutions and assimilated all peoples. Turner suggested that in the process of westward expansion, Euro-Americans on the frontier transformed themselves into a new homogeneous "American" character, infused with the spirit of ambition, innovation, and democracy. Turner wrote: "In the crucible of the frontier the immigrants were Americanized, liberated, and fused into a mixed race, English in neither nationality nor characteristics." Because of their frontier experience they left behind their Old World cultural traits and embraced an American democracy with innovation and self-reliance.[13] Contrary to this Turnerian view, other historians note that the frontier provided an escape and a space for the persistence of immigrant cultural traditions, suggesting that Turner simply ignored this element of immigrant frontier experience. The traditional Turnerian frontier historiography has largely obscured the history of many immigrant groups, other than to say that the frontier served to Americanize them.[14] The Basques represent a good example of an immigrant group whose history in the United States has been obscured by Turner's theory of the frontier process.

Before Frederick Jackson Turner expounded his frontier thesis in Chicago, contemporary Theodore Roosevelt had finished his first two volumes of *The Winning of the West* (1889).[15] In the first volume, the future president of the United States was more explicit in the interaction between "race" and environment in the western conquest. He looked at the "successful settlement" of the western territories by the Anglo-Saxon "race" and saw a different settlement pattern from most other major European colonial powers. Despite the interaction of groups from diverse European racial and ethnic backgrounds, Roosevelt argued that the "English-speaking race" prevailed and consolidated upon others and initiated a process of assimilation to the culture of the dominant Anglo-Saxon population. "Rival European races" and Indians, Roosevelt said, were accommodated to the dominant culture, but in the case of the Indians, exclusion occurred.[16]

Theodore Roosevelt made a brief comparison between the expansion of the United States across the continent and the extension of the Roman Empire in the Mediterranean.[17] He noted that the Roman Empire occupied spaces where Bretons, Celts, and Basques lived and in many instances these groups retained their ancient cultures, adapting to the power and the innovation of the Romans. Roosevelt, however, like Turner, saw the power of the American experience (what Turner would have called frontier process) as an important force in the complete "Americanization" of the European immigrants beyond simply the pale of the "English-speaking peoples."

Unlike the American experience, Roosevelt observed that in Europe, and even Latin America in the face of European colonial expansion, pockets of native cultures remained unassimilated. Roosevelt wrote: "Moreover, exactly as in Europe little ethnic islands of Breton and Basque stock have remained unaffected by the Romance flood, so in America there are large communities where the inhabitants keep unchanged the speech and the customs of their Indian forefathers."[18]

Roosevelt unexpectedly drew attention to the Basques as an ancient European unconquered "race." Frederick Jackson Turner read Roosevelt's work and reviewed *The Winning of the West* for the *Dial* magazine, praising it highly. In 1893, Turner declared the American frontier closed based on the census of 1890. In effect, Turner concluded that frontier expansion had been successfully completed, confirming Roosevelt's interpretation of the victorious settlement of the Anglo-Saxon "race" across the American continent.[19]

Although Theodore Roosevelt did not refer to the Basque immigrant sheepherders in the West, he viewed Basque people as an old European racial stock and unmixed race of ancient lineage that would foster an acceptable ethnicity—or cultural identity—to Euro-American cultural standards. The late nineteenth-century European ethnoracial theoretical conceptions about the Basques "museumized" everything related to them, including their representations in the American West.[20] This perception would eventually intermingle with the romanticized Turnerian frontier myth further championing the opportunities of the "free land" frontier for those with the grit and determination to settle it, as well as the racialized white Americans' views.

Starting in the early twentieth century, initial scholarship on Basque immigration in the United States tended to emphasize the rapid assimilation of Basque immigrants into the American mainstream. The central assumption of these early studies was that Basque immigrants were expected to adapt their ways easily and quickly because of their assumed docile and industrial race. These early studies tended to frame Basques as industrious people, members of a homogenous racial group coming from a homogenous country and culture, who were easily assimilable, and were quickly assimilated while retaining some of their cultural traits. Furthermore, conventional narratives have often argued that Basques were "racially" and culturally preadapted to sheep grazing conditions before their arrival in the American West.[21] It is a narrative, which—as I demonstrate throughout this book—can be clearly seen in the correspondence and ideas of ranchers, senators, congressmen, and others associated with the sheepherding industry throughout the nineteenth and twentieth centuries.

Pierre Lhande's 1910 book *L'Emigration Basque* ("The Basque Emigration"), which has served as a foundational text of the Basque diaspora scholarship, emphasized how Basque emigrants anywhere, under different conditions, were able to adapt easily to the host society's standards and even become "good" citizens wherever they immigrated. He related Basque emigration to features of the Basque society itself, with particular focus on the stem family structure with its inheritance system. According to Lhande, emigration was an inherent trait of Basque people who had an atavistic need—even "anxiety"—to emigrate.[22] From this assimilationist perspective, Basque immigrants in the American West easily adapted themselves to the sheepherding occupation in open-range conditions with which Basques were unfamiliar.

In the United States, at the same time, the good Basque sheepherder image was manufactured by public opinion and especially by those ranchers interested in employing additional Basque immigrants to work in their operations. Some American observers from the early twentieth century also embraced the idea that Basque immigrant sheep workers were racially qualified to tend sheep under extreme conditions of isolation, physical hazards, and risks of inclement weather. In 1917, Sol Silen in his *La historia de los Vascongados en el oeste de los Estados Unidos* ("The History of the Basques in the American West") described Basque immigrants as members of an "admirable race," whose main virtues were friendliness, honesty, truthfulness, and loyalty. The immigrants' work herding sheep flocks in the desolate rangelands, far from the employers' supervision, according to Silen, was an example of their honesty to the sheep owner.[23] Such distorted accounts about the Basque sheepherders were common in the early twentieth century, and they persisted and echoed within the academy.

American intellectuals, generally, tended to define the Basque people inhabiting the West in a racial and physical way. Racial depictions of the Basques almost always implied a good judgment, which in turn would bring an inevitable idealization of the Basque sheepherders in the context of a mythicized western frontier environment. They erroneously said that Basques settled in the Interior West and entered the sheepherder occupation because they had previous experience under similar conditions. In the mid-1920s, Ione B. Harkness wrote that Basques went to "seek in the new world a location similar in topography and climate and adapted to their ancient occupation of sheep herding."[24]

In the late 1930s and 1940s, various sociologists came to analyze the Basque immigrant social experience in the American West, with a special focus on their cultural transfer, social change, and adaptation in their new American context.[25] The methodological preoccupations of those sociologists were primarily the assimilation process of the Basque community into the American mainstream culture, paying special attention to the speedy "Americanization" of Basque immigrants. They contended that Basques accustomed themselves to the new environment, became an active social force, and fully integrated to host societies all around the West. They spoke of the Basques' "complete amalgamation," as Mary S. Wilcox said, and suggested that those immigrants easily adapted and assimilated in America.[26] Furthermore, as Basques settled in sparsely populated or isolated rural areas, these scholars observed, their integration process was constrained by the slow pace. Although they conflated Basque people with a "race" and nurtured the late nineteenth-century racialist theories that adhered to ideas of peculiar racial traits, some acknowledged that Basques arrived in America with a lack of background knowledge in sheep agriculture and that only, "accidentally," did they adapt to sheepherding.[27]

In the 1960s and 1970s, just as the Basque-American community gained visibility and this longtime immigrant flow began to decrease in the West, some new scholars analyzed the Basque immigrants' social assimilation, paying a special attention to the process of interaction between the immigrants and the host society.[28] Contrary to the idea that Basques were more easily integrated into the American society, various scholars revealed how in reality Basque immigrants faced considerable discrimination and exclusion in the early twentieth century, like many other immigrant groups, following a trajectory from a negative to a positive image.[29] Just as ethnicity replaced race as the defining category of Basques, new methodologies and interdisciplinary interaction brought new insights into the understanding of the Basque immigrant experience in the West. In 1970, anthropologist William A. Douglass observed that Basque immigrants, coming mostly from a peasant background, had no prior experience working as sheepherders under extensive open-range conditions.[30]

Later, some historians have put the Basque immigrant experience into a broader American context. In 1991, Richard W. Etulain discussed the tensions between the Basque immigrant community and the dominant host society. Etulain came to explain that in the late nineteenth and early twentieth centuries, both negative and positive images toward Basque immigrants coexisted depending on the specific issues at stake. It can also be concluded from his work that as some early Basque immigrants became prominent ranchers and second-generation Basque Americans integrated into American mainstream, Basques gradually obtained a good reputation among their neighbors in the different local communities in the West.[31]

Richard Etulain's work offered a new perspective and opened new avenues for understanding the Basque immigrant experience in the American West. After him, other experts on Basque immigration in the United States —such as Jeronima Echeverria, John Bieter, or Kevin D. Hatfield—have also adopted a revisionist view, which is the main point of departure for the present study.[32] Kevin D. Hatfield has correctly said, "The absence of a concerted examination of the social, political, and economic evolution of Basque open-range sheepherding in the American West has allowed the romantic ideal of the nomadic Basque shepherd to remain entrenched in popular literature."[33]

▲ ▲ ▲

An important aim of this book is to take a step toward filling this gap in the new historiography of the Basque in the American West. Echoing the New Western historians, the present study considers the West a complex place of interactions, convergences, and reciprocal influences among native-born and immigrant groups. Among the latter were also the Basques. As one historian has said, "Modern America is uniquely, as Turner failed to perceive, a blend of its immigrant and native heritages."[34] This book challenges the idea of melting-pot homogeneity in the American West by examining the Basque immigrant experience in Nevada as a complex collectivity in a complex geographical place.[35] The present study argues that the story is more complicated if the lives of the Basque immigrant workers are considered in their larger cultural dimensions. Furthermore, in the same vein as Kevin Hatfield's work, this book revises the traditional interpretation of the romantic image that surrounds Basque sheepherders in the American West.

The scholarship focusing on Basque immigration to the West has lacked a solid focus on the labor and political history of the Basque involvement in the sheepherding industry. In early works on the Basque immigrant communities, a romanticized perspective on the Basque sheepworkers was frequently maintained without engaging fully in the social, economic, or political history of this immigrant community. Contrary to the conventional and filiopietistic historical narrative, this book provides a critical study of the Basque sheepherders focused on understanding their place in its wider economic, political, sociological, and international context. Deconstructing the archetypal image of the Basque sheepherders in the West, this study fully integrates the Basque immigrant labor into the agricultural production system. The experience of Basque immigrant laborers was not isolated. It was vinculated with all the elements that made up the sheep industry, which was at the same time intermingled with broader capitalist relations of production through its many overlapping transformations on the American West during the late nineteenth and twentieth centuries.

In particular, this book explores the case of Basque agricultural labor in Nevada by integrating recent historiographical approaches that emphasize the impact of capitalism on the environment of the American West and vice versa. Historian William Cronon has argued: "Each of the city's commodities had been produced by human beings facing each other in the tumultuous relationship whose name was market: farmers and grain traders, cowboys and cattle barons, lumberjacks and lumbermen, all struggling over who would control the product of their collective work."[36]

Yet, as Cronon remarked, "The buying and selling of wage labor was among the most important innovations" that resulted from the establishment of capitalist modes of production in the West.[37] Based on this theoretical framework, this study considers Basque immigrant workers as another labor commodity in the global economic relations and fluctuations of the commodity market. In other words, Basque immigrant labor in the sheep industry must be comprehended as an interdependent element in the broader socioeconomic system.

Another aim of the present book is to understand the racialization of the Basque immigrants and how the racial meaning influenced the integration process of this collectivity to the host society. In the early 1970s, William Douglass identified the Basque immigrant population as Caucasian and took for granted the privileges bestowed to them under this racial category in the United States. He wrote: "The fact that the Basques are Caucasians meant that they did not possess physical clues which would make them readily identifiable by the uninformed and hence easy marks for discrimination on racist grounds."[38] However, the question here is not, as Douglass put it, about whether Basques were Caucasians or not; it is rather about who categorized—and why—Basques as Caucasians. Historian Matthew F. Jacobson has stated: "Caucasians are not born... they are somehow made. It's just a question of who does the making."[39]

Enriching to the preexisting historiography of "whiteness" in America, the present investigation considers "race" as an invented category, which has been founded politically and culturally in order to separate people based on supposed differences in the United States. Historians such as David R. Roediger, James R. Barrett, and Mathew F. Jacobson have investigated how race ascribed European immigrant groups a social and economic status crucial in order to accelerate their integration processes in the United States.[40] Moreover, according to Matthew F. Jacobson, those European immigrants, such as the Basques, could be identified as white, but at the same time they could be seen by others (or even by themselves) as racially different, even nonwhite, with all forms of discrimination and exclusion that this implied. The passage of the early 1920s immigration restriction laws, Jacobson has further said, resulted in the creation of an amorphous group called Caucasians in which the new immigrants from southern and eastern Europe were now included.[41]

Although immigration restrictive legislation limited significantly Basque immigration to the American West, after that their integration process accelerated amidst a shifting new racial paradigm and a rapidly changing international context marked by overlapping events leading up to World War II. During the shortage of Basque immigrant labor supply that ensued as a result of the early 1920s restrictive measures, Mexican labor was employed in the sheep operations, something that increased especially during World War II. During the war, sheep ranchers enthusiastically glorified and praised Basque sheepherders, and discriminated against Mexicans as undesired labor. The "Othering" and racist exclusion of Mexican sheepherders contributed further to more positive attitudes toward Basque immigrants, all of which must be understood in terms of whiteness in America.

This book challenges the essentialist conception of Basques by questioning the underlying assumption that they were naturally good sheep laborers.[42] Just as Basque immigrants were regarded as indispensable laborers in the western open-range sheep industry throughout the twentieth century, as I will show throughout the book, a Nevada woolgrower community manufactured a strand of a Basque identity on racial terms as a basic commodity, which had important repercussions in other social, economic, political, and international dimensions. This study suggests that the contemporary racist ideology led to the social construction of Basques as "good sheepherders," which strengthened and persisted in livestock agricultural circles. Then, the Basques effectively reached the point of visibility that would eventually grant them social legitimacy in the American West.

This book also aims to contribute to the existing historiography of American immigration from the 1920s to 1960s—a period characterized by major immigration restrictions in American history. This research builds primarily upon the work of historian Mae M. Ngai who has discussed the legal and juridical process of categorization of immigrants based on ethno-racial identification of the early 1920s that attempted to redefine the racial-ethnic hierarchical map of the United States, which is to say that all nonwhites were excluded from legal immigration. According to Ngai, "Euro-American identities turned both on ethnicity—that is, a nationality-based cultural identity that is defined as capable of transformation and assimilation—and on a racial identity defined by whiteness."[43] Drawing upon Ngai's work and the latest historiographical approaches of American immigration, the present study examines the impact and effect of the restrictive immigration laws upon the Basque emigration to the United States in the mid-twentieth century.

Furthermore, the present book explores the political and legal dimension of the Basque immigrant workers in the sheep industry. This book explores the subject of land use disputes against the Basques and other newcomers in Nevada, and the political attention it attracted from pro-cattle interests. Among them, the most vociferous was Senator Key Pittman, who in the early 1910s launched a campaign against the open-range sheep industry in Nevada whose targets included the Basque agricultural labor. The present book argues that the criminalization of Basque immigrant sheepherders and the denigration of the entire collectivity operated as an instrument to defend the economic interests of important cattle ranchers in Nevada. These assumed perceptions about the Basque sheepherders eventually influenced federal land management decisions. More specifically, in the early twentieth century the Forest Service found the problem of overstocking more troublesome, and according to the rangers, itinerant sheepherders were a major problem affecting Nevada's rangelands. The elite discourse about conservation villainized and deemed threatening the Basque sheepherders to vast public grazing areas in Nevada, which shaped American public opinion.

A major topic of the present study is the way ranchers responded to the Basque labor shortage and other economic problems affecting the sheep industry of Nevada from World War II to the mid-1950s, problems that ultimately caused the Basque sheepherders to become a focus for officials at the highest levels of the American and Spanish governments. This book analyzes the political organization of Nevada interest groups who sought to open the doors for an increasing Basque immigration from Spain in the mid-twentieth century in close cooperation with other western woolgrower associations under the patronage of Senator Patrick McCarran.[44]

The history of Pat McCarran and his lobbying efforts to recruit Basque immigrant laborers is at the same time the history of the United States' foreign relations with Francisco Franco's Spain during the forties and fifties. With the outbreak of the Cold War, Franco's regime deliberately used its fervent anticommunism to approach the United States for economic support. The United States for its part found Spain's geostrategic position attractive during the early Cold War. In this context, diplomatic relations facilitated economic relations, including the recruitment of Basque immigrants to work in the western sheep industry.

This book brings into focus this dimension of the United States–Spanish relationships under the Cold War dynamics through an examination of the immigration issues between the two governments.[45] This study understands the recruitment of Basque immigrants to work in the Nevada and western sheep industry as part of the international relations between the United States and Franco's Spain that eventually led to the Pact of Madrid of 1953.

▲ ▲ ▲

In order to clarify the major epochs surrounding this social, economic and political history, this book is broken out into three separate parts. The first part discusses the historical foundations for Basque immigration in

Nevada and the West. The first chapter traces the late nineteenth-century development of the open-range sheep industry in the state of Nevada, with special focus on its economic and legal dimension. Particularly, it explores how the free access to the public-domain lands and some favorable environmental conditions fostered the expansion of sheep grazing, which would eventually became a major pull factor for subsequent Basque immigration.

The second chapter analyzes the arrival of early Basque ranch interests in Nevada and their role in attracting future Basque immigrants from the Old Country to become sheepherders. The expansion of sheep agriculture in the 1890s brought the employment of significant numbers of workers, largely European immigrants, to labor in the most desolate rangelands of the Great Basin. During the 1890s, there was demand for contract labor from sheep operations, some of which were owned by Basques. Early Basque pioneer stock operators began recruiting Basque workers from the Old World. The efforts of these early contractors set the immigration pattern between the Basque Country and the American West.

The second part of the book is concerned with the decisive years when the Basque immigrant sheepherders became a noticed collectivity and further consolidated their place within the host society. The third chapter explores how at the turn of the century the Basques' presence in Nevada and the vast herds of sheep they tended influenced the early Forest Service decisions about who should graze and have access to high-mountain pastures. The Basque sheep herds also prompted actions from the state of Nevada and other local interests to put in place restrictions and rules on access to water and land by itinerant graziers.

Chapter 4 analyzes the historical process of social and economic adjustments of the Basque immigrant community between the decades of 1910 and 1930. By 1910, the steady flow of immigrants from the Basque Country guaranteed and strengthened the endurance of this community in Nevada. As they became a well-accepted and respected immigrant group, Basques began facing greater legal constraints on their status as immigrants and on their engagement in itinerant sheep grazing. The Immigration Act of 1924 sharply reduced Basque and other immigration from southern and eastern Europe. For the sheep industry, this meant a Basque labor shortage in the near future.

The third part of the book deals with the recruiting process of Basque immigrant workers that responded to the labor shortage during

World War II, a process that was prolonged until the 1950s and beyond. Chapters 5 and 6 analyze how sheep ranchers demanded from their representatives in Congress that legal restrictions to enter the country on Basque immigrants be lifted during World War II. According to them, the depletion of Basque labor to work on the ranges threatened to obstruct sheep production. They organized themselves to find solutions. In this fight for Basque sheepherders, Nevada's U.S. senator Patrick McCarran was instrumental in helping woolgrowers gain an increased Basque immigrant labor supply. Although it was not the only issue facing the sheep industry in the West, it drew great attention.

Chapter 7 sets the Basque immigrant labor issue in the context of diplomatic negotiations between the United States and Spain that eventually crystallized into the Pact of Madrid in 1953—the military-related cooperation contract between the two governments that officially ended the ostracism against Spain from western liberal democracies. This chapter analyzes how the process of importing more Basque immigrant labor from Spain was closely intertwined with the evolution of U.S.–Spanish relations during the early Cold War. It explores the interests of both Patrick McCarran and Francisco Franco in encouraging further Basque immigration to the American West.

In summing up, this book gives a more complex perspective on the Basque sheepherders in Nevada's Great Basin. The present study contributes to a detailed comprehension of the social, economic, political, and cultural world that surrounded Basque immigrant sheepherders from the late nineteenth to the mid-twentieth centuries in Nevada. It analyzes the place, participation, and negotiation of the Basque immigrant agricultural workers within the sheep industry of Nevada. On the one hand, it examines the place of the Basque sheepherders within the overall sheep business, locally, nationally, and internationally. On the other hand, it analyzes their participation in the development of the open-range sheep industry, as well as their civic participation in the host society. And further, it explores the complex identity negotiations that Basque immigrants faced at different times in the American context. This is a story of how Basque immigrants became sheepherders in Nevada and eventually were commodified as a desirable workforce in response to the constant demand for their labor power in the sheep industry from the late nineteenth to the mid-twentieth centuries.

Notes

1. "Oregon's Basques: A Friendly, Hard-Working Race," *Oregonian*, June 20, 1937, 10.
2. Ibid.
3. William A. Douglass, "Basque-American Identity: Past Perspectives and Future Prospects," in *Change in the American West: Exploring the Human Dimension*, ed. Stephen Tchudi (Reno: University of Nevada Press, 1996), 183–98; William A. Douglass and Jon Bilbao, *Amerikanuak: Basques in the New World* (Reno: University of Nevada Press, 1975).
4. The Basque presence in the North American West dates back to the Spanish colonial period in the American Southwest. Between the late sixteenth and early nineteenth centuries, Basques were directly involved in the Spanish Empire's expansion from colonial Mexico into territory north of the Rio Grande. The sustained inflow of Basque immigrants into the Far West, however, took place in the mid-nineteenth century during and after the California Gold Rush. For a detailed exposition on the earliest Basque presence in the American West, see: Douglass and Bilbao, *Amerikanuak*; Marc Simmons, *The Last Conquistador: Juan de Oñate and the Settling of the Far Southwest* (Norman: University of Oklahoma Press, 1991); Donald T. Garate, "Basque Names, Nobility, and Ethnicity on the Spanish Frontier," *Colonial Latin American Historical Review* 2, no. 1 (Winter 1993): 77–104; Donald T. Garate, "Juan de Oñate's *Prueba de Caballero*, 1625: A Look at His Ancestral Heritage," *Colonial Latin American Historical Review* 7, no. 2 (Spring 1998): 129–73; José Manuel Azcona Pastor, *Possible Paradises: Basque Emigration to Latin America* (Reno: University of Nevada Press, 2004); Donald T. Garate, *Juan Bautista de Anza: Basque Explorer in the New World, 1693–1740* (Reno: University of Nevada Press, 2005). See also: David J. Weber, *The Spanish Frontier in North America* (New Haven, Conn.: Yale University Press, 1992); David Goodman, *Spanish Naval Power, 1589–1665: Reconstruction and Defeat* (Cambridge and New York: Cambridge University Press, 1997); John L. Kessell, *Spain in the Southwest: A Narrative History of Colonial New Mexico, Arizona, Texas and California* (Norman: University of Oklahoma Press, 2002).
5. On the concept of "immigrant enclave," see: Alejandro Portes and Robert D. Manning, "The Immigrant Enclave: Theory and Empirical Examples," in *Competitive Ethnic Relations*, eds. Susan Olzak and Joane Nagel (Orlando, FL: Academic Press, 1986), 47–68. See also: Ivan Light, "Immigrant and Ethnic Enterprise in North America," *Ethnic and Racial Studies* 7, no. 2 (April 1984): 195–216.
6. The term "chain migration" refers to the process by which migrants follow the path of other kin or former neighbors who had emigrated earlier, and joined them to a specific destination. See: Monica Boyd, "Family and Personal Networks in International Migration: Recent Developments and New Agendas," *International Migration Review* 23, no. 3 (Fall 1989): 638–70. See also: Roger Daniels, *Coming to America: A History of Immigration and Ethnicity in American Life* (New York: Perennial, 2002).
7. On ethnic occupational mobility in general, see: William L. Yancey, Eugene P. Ericksen, and Richard N. Juliani, "Emergent Ethnicity: A Review and Reformulation," *American Sociological Review* 41, no. 3 (June 1976): 391–403.

8. This study applies the concept "racialised" following the work of sociologist Stephen Small, *Racialised Barriers: The Black Experience in the USA and England in the 1980s* (London and New York: Routledge, 1994); Small, "The Contours of Racialisation: Structures, Representations and Resistance in the USA," in *Race, Identity and Citizenship: A Reader*, eds. Rodolfo D. Torres, Louis F. Mirón, and Jonathan Xavier Inda (Malden, Mass.: Blackwell, 1999).
9. Scholars have been writing about the "new immigration" from Europe, and yet not much has been investigated for the following years. Pertinent major works include: John Bodnar, *Immigration and Industrialization: Ethnicity in an American Mill Town, 1870–1940* (Pittsburgh: University of Pittsburgh Press, 1977); Virginia Y. McLaughlin, *Family and Community: Italian Immigrants in Buffalo, 1880–1930* (Urbana and Chicago: University of Illinois Press, 1977); Charles C. Moskos Jr., *Greek Americans: Struggle and Success* (Englewood Cliffs, N.J.: Prentice Hall, 1980); Elizabeth Ewen, *Immigrant Women in the Land of Dollars* (New York: Monthly Review Press, 1985); Ewa T. Morawska, *For Bread with Butter: Life-Worlds of East Central Europeans in Johnstown, Pennsylvania, 1890–1940* (New York: Cambridge University Press, 1985); John J. Bukowczyk, *And My Children Did Not Know Me: A History of the Polish-Americans* (Bloomington and Indianapolis: Indiana University Press, 1987).
10. At the same time, this study joins the recent scholarship on European immigration in the American West in analyzing the disparate Euro-immigrant workers and collectivities in this region and their divisions. See, for example: Robert C. Ostergren, *A Community Transplanted: The Trans-Atlantic Experience of a Swedish Immigrant Settlement in the Upper Middle West, 1835–1915* (Madison: University of Wisconsin Press, 1988); David M. Emmons, *The Butte Irish: Class and Ethnicity in an American Mining Town, 1875–1925* (Urbana: University of Illinois Press, 1989); Dino Cinel, *From Italy to San Francisco: The Immigrant Experience* (Stanford, Calif.: Stanford University Press, 1982); Anna Zellick, "'We All Intermingled:' The Childhood Memories of South Slavic Immigrants in Red Lodge and Bearcreek, Montana, 1904–1943," *Montana, The Magazine of Western History* 44, no. 3 (Summer 1994): 34–45; Rob Kroes, *The Persistence of Ethnicity: Dutch Calvinist Pioneers in Amsterdam, Montana* (Urbana: University of Illinois Press, 1992).
11. The following works are notable examples: Frederick C. Luebke, ed., *European Immigrants in the American West: Community Histories* (Albuquerque: University of New Mexico Press, 1998); Gunther Peck, *Reinventing Free Labor: Padrones and Immigrant Workers in the North American West, 1880–1930* (New York: Cambridge University Press, 2000); Jessie L. Embry and Brian Q. Cannon, eds., *Immigrants in the Far West: Historical Identities and Experiences* (Salt Lake City: University of Utah Press, 2015).
12. Few European ethnic groups have been more carelessly analyzed in the traditional historiographic narrative than the Basque immigrants. Anthropologist William Douglass has said that Basques are "one of the least-studied elements in the pluralistic social fabric of the Americas despite the fact that they were among the first Europeans to emigrate to the New World, as well as one of its most widely distributed immigrant groups." Douglass and Bilbao, *Amerikanuak*, 1.

13. Frederick Jackson Turner, *The Frontier in American History* (New York: Henry Holt and Company, 1948), 22–23.
14. Historian Frederick Luebke has observed the Turnerian mode "predisposed the historian to emphasize the ease and rapidity with which ethnic groups were assimilated into American society and to ignore ethnocultural conflict and the persistence of immigrant attitudes, values, and behaviors." Frederick C. Luebke, "Turnerism, Social History, and the Historiography of European Ethnic Groups in the United States," in *Germans in the New World: Essays in the History of Immigration*, ed. Frederick C. Luebke (Chicago: University of Illinois Press, 1990), 140.
15. Henry N. Smith, *Virgin Land: The American West as Symbol and Myth* (Cambridge, Mass.: Harvard University Press, 1950), 250–51.
16. Theodore Roosevelt, *The Winning of the West*, vol. 1 (New York: G.P. Putman's Sons, 1889), 13–17.
17. More recently, some scholars have seen broad historical analogies between the Roman Empire and the rise of the United States to world power. See, for example: Chalmers Johnson, *The Sorrows of Empire: Militarism, Secrecy, and the End of the Republic* (New York: Metropolitan Books, 2004); Niall Ferguson, *Colossus: The Price of America's Empire* (New York: Penguin Press, 2004).
18. Roosevelt, *Winning of the West*, 13.
19. Roosevelt, *Winning of the West*, 11–13; Frederick Jackson Turner, review of *The Winning of the West*, by Theodore Roosevelt, *The Dial* 10 (August 1889): 71–73; Richard Hofstadter, *The Progressive Historians: Turner, Beard, Parrington* (New York: Alfred A. Knopf, 1968), 47–61.
20. I have borrowed this term "museumized" from historian Patricia N. Limerick, *The Legacy of Conquest: The Unbroken Past of the American West* (New York: W.W. Norton, 1987), 25.
21. Some cultural geographers apply the concept of "preadaptation" to the transplantation and adjustment process by European colonizers from the Old to the New World. The success degree of European immigrants' settlement in America depended upon their strategy to select old economic systems and adopt new adaptive means in their new surroundings. Milton Newton, "Cultural Preadaptation and the Upland South," *Geoscience and Man* 5 (June 1974): 144; Terry G. Jordan, "Preadaptation and European Colonization in Rural North America," *Annals of the Association of American Geographers* 79, no. 4 (December 1989): 494. See also: Terry G. Jordan, "New Sweden's Role on the American Frontier: A Study in Cultural Preadaptation," *Geografiska Annaler* 71, no. 2 (1989): 71–83; Terry G. Jordan and Matti Kaups, *The American Backwoods Frontier: An Ethnic and Ecological Interpretation* (Baltimore, Md.: Johns Hopkins University Press, 1989).
22. Pierre Lhande, *L'Emigration Basque: Histoire, économie, psychologie* (Paris: Nouvelle Librairie Nationale, 1910), 11–13, 22–23, 155, 157.
23. Sol Silen, *La historia de los Vascongados en el oeste de los Estados Unidos* (New York: Las Novedades, 1917), 5, 6.
24. Ione B. Harkness, "Basque Settlement in Oregon," *Oregon Historical Quarterly*

34, no. 3 (September 1933): 273. See also: Ione B. Harkness, "Certain Community Settlements in Oregon" (master's thesis, University of Southern California, 1925).

25. Mary S. Wilcox, "A Historical Study of the Basque Race with Special Reference to the United States" (master's thesis, University of Utah, 1939); Joseph H. Gaiser, "The Basques of the Jordan Valley Area: A Study in Social Processes and Social Change" (PhD diss., University of Southern California, 1944); Flavia M. McCullough, "The Basques in the Northwest" (master's thesis, University of Portland, 1945); John B. Edlefsen, "A Sociological Study of the Basques of Southwest Idaho" (PhD diss., State College of Washington, 1948); Carol M. Pagliarulo, "Basques in Stockton: A Study of Assimilation" (master's thesis, College of the Pacific, 1948).
26. Wilcox, "A Historical Study of the Basque Race," 135.
27. Gaiser, "The Basques of the Jordan Valley Area," 42, 80.
28. Clifford A. Sather, "Marriage Patterns among the Basques of Shoshone, Idaho" (master's thesis, Reed College, 1961); Allura N. Ruiz, "The Basques-Sheepmen of the West" (master's thesis, University of Nevada, 1964); James P. Kelly, "The Settlement of Basques in the American West" (senior paper, Harvard University, May 1967); Grant E. McCall, "Basque-Americans and a Sequential Theory of Migration and Adaptation" (master's thesis, San Francisco State College, 1968); Joseph R. Castelli, "Basques in the Western United States: A Functional Approach to Determination of Cultural Presence in the Geographic Landscape" (PhD diss., University of Colorado, 1970); William A. Douglass, "The Basques of the American West: Preliminary Historical Perspectives," *Nevada Historical Society Quarterly* 13, no. 4 (Winter 1970): 12–25; John A. Stafford, "Basque Ethnohistory in Kern County, California: 1872–1934 A.D." (master's thesis, Sacramento State College, 1971); Louise B. Dunn, "The Salt Lake City Basque Community: Atypical in the American West" (master's thesis, University of Utah, 1972); Sarah C. Baker, "Basque American Folklore in Eastern Oregon" (master's thesis, University of California, Berkeley, 1972); William A. Douglass, "Lonely Lives under the Big Sky," *Natural History* 82, no. 3 (March 1973): 28–39; Richard H. Lane, "The Cultural Ecology of Sheep Nomadism: Northeastern Nevada, 1870–1972" (PhD diss., Yale University, 1974); Frank P. Araujo, "Basque Cultural Ecology and Echinococcosis in California" (PhD diss., University of California, Davis, 1974); Douglass and Bilbao, *Amerikanuak*; Mademoiselle Bouesnard, "Basque Emigration to California and Nevada since 1960" (master's thesis, Université de Pau et des Pays de l'Adour, 1976); Sonia J. Eagle, "Work and Play among the Basques of Southern California" (PhD diss., Purdue University, 1979); Jean F. Decroos, "The Long Journey: Assimilation and Ethnicity Maintenance among Urban Basques in Northern California" (PhD diss., University of Oregon, 1979). See also Daniel Alexander Gómez-Ibáñez, "The Rise and Decline of Transhumance in the United States" (master's thesis, University of Wisconsin, 1967).
29. Wilbur S. Shepperson, *Restless Strangers: Nevada's Immigrants and Their Interpreters* (Reno: University of Nevada Press, 1970), 41, 58–59.
30. Douglass, "The Basques of the American West," 19–20.

31. Richard W. Etulain, ed., *Basques of the Pacific Northwest* (Pocatello: Idaho State University Press, 1991), 7, 20, 80–90.
32. Jeronima Echeverria, *Home Away from Home: A History of Basque Boardinghouses* (Reno: University of Nevada Press, 1999); John Bieter and Mark Bieter, *An Enduring Legacy: The Story of Basques in Idaho* (Reno: University of Nevada Press, 2000); Kevin D. Hatfield, "'We Were Not Tramp Sheepmen': Resistance and Identity in the Oregon Basque Community, Accustomed Range Rights, and the Taylor Grazing Act, 1890–1955" (PhD diss., University of Oregon, 2003), 18, 27–28. Later, Hatfield published an article summarizing his thesis: Kevin D. Hatfield, "'We Were Not Tramp Sheepmen': Joe Odiaga and Acculturation, Resistance, and Identity in the Bizkaian Basque Community, 1890–1946," *Nevada Historical Society Quarterly* 52, no. 4 (Winter 2009): 292–312.
33. Hatfield, "'We Were Not Tramp Sheepmen,'" 2003, 18.
34. Robert M. Utley, *The Indian Frontier of the American West, 1846–1890* (Albuquerque: University of New Mexico Press, 1984), 272.
35. On the complex histories, geographies, and cultures of the American West, see, for example: David M. Wrobel and Michael C. Steiner, eds., *Many Wests: Place, Culture, and Regional Identity* (Lawrence: University Press of Kansas, 1997).
36. William Cronon, *Nature's Metropolis: Chicago and the Great West* (New York: W.W. Norton, 1991), 149.
37. Ibid. Along with Cronon's book, the following works are worth mentioning: David Vaught, *Cultivating California: Growers, Specialty Crops, and Labor, 1875–1920* (Baltimore, Md.: Johns Hopkins University Press, 1999); David Igler, *Industrial Cowboys: Miller & Lux and the Transformation of the Far West, 1850–1920* (Berkeley: University of California Press, 2001); Nancy Langston, *Where Land and Water Meet: A Western Landscape Transformed* (Seattle and London: University of Washington Press, 2003).
38. Douglass, "Basques of the American West," 16. Later, Douglass developed the same thesis in his book (coauthored with Jon Bilbao), *Amerikanuak*, 3.
39. Mathew F. Jacobson, *Whiteness of a Different Color: European Immigrants and the Alchemy of Race* (Cambridge, Mass.: Harvard University Press, 1998), 3.
40. David R. Roediger, *The Wages of Whiteness: Race and the Making of the American Working Class* (London and New York: Verso, 1991); James R. Barrett and David Roediger, "How White People Became White," in *Critical White Studies: Looking Behind the Mirror*, eds. Richard Delgado and Jean Stefancic (Philadelphia: Temple University Press, 1997); Jacobson, *Whiteness of a Different Color*.
41. Jacobson, *Whiteness of a Different Color*, 91–96.
42. See "essentialism" in Alan Barnard and Jonathan Spencer, eds., *Encyclopedia of Social and Cultural Anthropology* (London and New York: Routledge, 1996), 188–190.
43. Mae M. Ngai, *Impossible Subjects: Illegal Aliens and the Making of Modern America* (Princeton, N.J.: Princeton University Press, 2004), 3, 7; for the historiography of American immigration, this study also is primarily based on the following scholarly books: Keith Fitzgerald, *The Face of the Nation: Immigration, the State, and the National Identity* (Stanford, Calif.: Stanford University Press, 1996); Desmond S.

King, *Making Americans: Immigration, Race, and the Origins of Diverse Democracy* (Cambridge, Mass.: Harvard University Press, 2000); Aristide R. Zolberg, *A Nation by Design: Immigration Policy in the Fashioning of America* (Cambridge, Mass.: Harvard University Press with Russell Sage Foundation, 2006).

44. This book is intended to complement the various scholarly literature of Patrick McCarran's political career by analyzing his battle for the sake of the Basque immigrant workers. See particularly the following works: Margaret P. McCarran, "Patrick Anthony McCarran, 1876–1954," Part I, *Nevada Historical Society Quarterly* 11, no. 3–4 (Fall/Winter 1968): 5–66; Margaret P. McCarran, "Patrick Anthony McCarran, 1876–1954," Part 2, *Nevada Historical Society Quarterly* 12, no. 1 (Spring 1969): 5–75; Von V. Pittman Jr., "Senator Patrick A. McCarran and the Politics of Containment" (PhD diss., University of Georgia, 1979); Jerome E. Edwards, *Pat McCarran: Political Boss of Nevada* (Reno: University of Nevada Press, 1982); Harold G. Rader, "The McCarran Doctrine: An Abrogation of Federal Sovereign immunity through the McCarran Adjudication Amendment" (master's thesis, University of Nevada, Reno, 1993); Michael J. Ybarra, *Washington Gone Crazy: Senator Pat McCarran and the Great American Communist Hunt* (Hanover, N.H.: Steerforth Press, 2004).

45. This study complements the existing scholarship dealing with the diplomatic relations between the United States and Franco's Spain. This recent scholarship adopts the basic assumption that in reality Franco's regime was not so much isolated from the Western world during the forties and fifties. See the following works: Ángel Viñas, *Los pactos secretos de Franco con Estados Unidos: Bases, ayuda económica y recortes de soberanía* (Barcelona: Grijalbo, 1981); Boris N. Liedtke, *Embracing a Dictatorship: US Relations with Spain, 1945–53* (New York: St. Martin's Press, 1998); Jill Edwards, *Anglo-American Relations and the Franco Question, 1945–1955* (Oxford: Clarendon Press, 1999); Aurora Bosch and María Fernanda del Rincón, "Dreams in a Dictatorship: Hollywood and Franco's Spain, 1936–1956," in *"Here, There and Everywhere": The Foreign Policy of American Popular Culture*, eds. Reinghold Wagnleitner and Elaine Tyler May (Hanover, N.H.: University Press of New England, 2000); Oscar Calvo González, "¡Bienvenido, Míster Marshall! La ayuda económica americana y la economía española en la década de 1950," *Revista de Historia Económica* 19 (2001): 253–76; Lorenzo Delgado and María Dolores Elizalde, eds., *España y Estados Unidos en el siglo XX* (Madrid: CSIC, 2005); Pablo León Aguinaga, *Sospechosos habituales: El cine norteamericano, Estados Unidos y la España franquista, 1939–1960* (Madrid: Consejo Superior de Investigaciones Científicas, 2010); Joan Maria Thomàs, *Roosevelt, Franco, and the End of the Second World War* (New York: Palgrave Macmillan, 2011); Antonio Niño Rodríguez and José Antonio Montero Jiménez, eds., *Guerra fría y propaganda: Estados Unidos y su cruzada cultural en Europa y América Latina* (Madrid: Biblioteca Nueva, 2012).

PART I

After the Sheep Rush

1

The Promises of the Silver State

The Development of the Sheep Industry in Nevada, 1850–1900

Any investigation of Basque immigrant labor in Nevada's sheep industry forms a small part of the saga of the western livestock range history. In this particular study, the focus is the socioeconomic development of commercial sheep grazing in the Nevada portion of the Great Basin during the late nineteenth and early twentieth centuries.[1] However, before the arrival of Basque immigrants to this region, there was already a sheep industry well underway in Nevada. Before a consideration of Basque immigration and work as sheepherders can be undertaken, there must occur a consideration of the sheep industry itself in Nevada during the last half of the nineteenth century.

Nevada is an arid, mountainous, and barren state. It consists mainly of a plateau marked by north to south mountain ranges lying for the most part within the Great Basin desert. The precipitation is low and consists for the most part of snow falling in the high mountains during winter. The precipitation at lower elevations is too low for even dry farming purposes. Because of this arid ecosystem, Nevada has been unsuited to most kinds of crop agriculture, and consequently the state's population remained low and sparse in the nineteenth century. During the early mining boom in the mid-nineteenth century, the population was concentrated in small urban mining developments. By the end of the nineteenth century, mining fell into a depression in Nevada. Then, livestock operations surpassed a failing mining industry in the state. The scarcity of water sources determined much of the socioeconomical activity of Nevada. Throughout the years, the possession and control of water determined the use of Nevada's rangelands.[2]

The mining industry was the first modern extractive sector to be developed by the Euro-American population in Nevada. The continued arrival of newcomers who tried to prosper in the so-called Silver State in the extraction of minerals added to the development of the livestock industry. In other words, miners' and early settlers' mouths needed feeding. After the Civil War (1861–65), Nevada's rangelands supported thousands of cattle and, to a lesser degree, sheep. By the late 1880s, the sheep industry became a fast-growing sector. For many with capital to invest in the American West, sheep grazing was a profitable business, offering an opportunity for profits with low investment for the cost of production. In other words, the sheep industry appeared to be a low-risk business.

In the late nineteenth century, the state of Nevada became one of the major sheep producing states in the West, owing largely to the presence of a very high percentage of public-domain lands. Free access to these public-domain lands was important in allowing the sheep business to be profitable in Nevada. In the late 1880s, the state of Nevada was regarded as "the best sheep country" in the United States.[3] In the 1890s, as the mining camps declined, sheep numbers multiplied on Nevada ranges. However, sheep numbers would fluctuate over the coming decades during policy debates and rules changes regarding the use of the free lands.[4]

During this era, Basque immigration to the region increased substantially as work opportunities in the sheepherding industry became available (the way this immigration unfolded is discussed in detail in chapter 2). In chapter 1 below, I discuss the historical background of the sheep industry in the late nineteenth century in order to bring into view the broader historical context of Basque immigrant sheepherding in this region.

"Through Nevada's Desert": Commercial Sheep Grazing Begins

The economic evolution of the Nevada Territory and early statehood period was largely determined by the development of the Far West's mining economy. As the demand for animal products increased, the sheep industry became a highly profitable business. The first domestic sheep entering Nevada was a flock introduced by the Workman and Rowland party in 1841 who drove a herd of 150 sheep from New Mexico to California, passing through the southern portion of the present boundaries of

Nevada. The Workman-Rowland migration preceded subsequent larger sheep drives from New Mexico to California in the early 1850s.[5]

During the gold rush in California in 1849 and the early 1850s, Nevada emerged as an important site within the western sheep market. Generally, the point of departure for the sheep herds into California became New Mexico, where investors could find sheep at low prices. In the long livestock drives, thousands of head of sheep moved from northwestern New Mexico to the mining camps in California. Nevada became part of this trail on the way to California. In 1852, one of the major, if not first, totally commercial sheep drives to California was Richens Lacy Wootton, also known as "Uncle Dick," who trailed one thousand sheep from Taos in New Mexico to Sacramento with the help of fourteen hired sheepherders, seven guards, eight goats, and sheep dogs to control the flocks. One year later, Christopher "Kit" Carson, along with his partner Lucien B. Maxwell, bought thirteen thousand head of sheep in Santa Fe and trailed them to Sacramento. Both parties inspired later sheep drives and established the route for the trailing of these animals to the Pacific Coast. These early entrepreneurs made rapid fortunes, later carrying out additional successful sheep drives across the region. One source calculated that between 1852 and 1860 more than half a million sheep traversed Nevada on their way to California.[6]

Livestock agriculture had its beginnings in Nevada in the 1850s in the western portion of the Great Basin. Family-based Mormon migrants from Utah began to settle in the Carson Valley area and developed rural communities, bringing with them small flocks of sheep. Also, stockmen from California began wintering their cattle in the western rangelands of Nevada, primarily in the Truckee Meadows, as forerunners of the eastward movement of the California ranch frontier. By the second half of the 1850s, three prominent livestock ranching operations had already settled in Nevada: Fred Dangberg in Carson Valley, Hock Mason in the area of today's Lyon County, and the Smith Brothers in the area that became known by their family name, Smith Valley.[7]

At the same time, the decline of placer mining in California prompted prospectors and capitalists to explore for new precious metals in the interior of the Far West. New lodes were found immediately across the Sierra. In 1859, the discovery of the Comstock Lode opened a new prosperous local economy and, of course, strengthened the small livestock industry east of the Sierra in what was then Utah Territory. Stock operators prospered and succeeded along with the mining bonanzas of the Comstock and continued even after the failure of the Comstock mines around 1880. After the initial Comstock discoveries, California corporations rushed to purchase mining claims in the Comstock Lode, something which was followed by the quick development of the San Francisco Mining Stock Exchange to raise investment capital. Subsequent mining booms throughout Nevada increased demand for livestock products.[8]

Although the relation between mining discoveries and a livestock economy was important in the development of the livestock industry in Nevada, it does not tell the whole story as the livestock economy continued to also be viable during the twenty years of Nevada's mining depression from 1880 to 1900. Nevertheless, the success of early stockmen during the mining era helped prove the potential of selected Nevada ranges for grazing. Some of those early operations included cattle herds from Texas which, in the years immediately following the Civil War, began to be grazed in Elko County, and then in the early 1870s, the important operations established by several prominent ranchers coming from California and elsewhere, such as John W. Long's operation in Clover Valley and the Altube Brothers in Independence Valley. Later, in 1890, historian Hubert H. Bancroft would optimistically write: "Nevada is a better agricultural country than at first glance one might expect to find."[9]

After Nevada statehood in 1864 and the end of the Civil War in 1865, the future for the stock industry on the open rangelands of Nevada had been promising. Expansion of livestock production in the Pacific Coast brought falling stock market prices in California. Large sheep bands now grazed eastward into the Great Basin to serve the satellite markets created around the new Nevada mining districts. In the spring of 1869, the Central Pacific Railroad completed its route across Nevada. The railroad set the stage for additional market access to livestock raised on Nevada ranges. New mineral discoveries increased the railroad traffic, as did livestock production. Along the railroad's route, property values significantly rose and stock agriculture in adjoining counties expanded. In the summer of 1872, the *Humboldt Register* reported that an increasing number of "sheepmen" were acquiring titles to lands in Humboldt County, Nevada. Because sheep products were more valued in Chicago and other far-away markets on the East Coast, the development and improvement of transportation and favorable market conditions made Nevada a perfect state to raise meat on the hoof. Free forage from federal lands, of course, lowered production costs. All contributed to the ultimate objective of securing the largest margin of profit.[10]

As the mining industry declined in Nevada in the latter part of the century, the livestock industry continued to grow. In 1875, the *Territorial Enterprise* of Virginia City announced that Nevada did not need more miners and advised those migrants who intended to be employed by those corporations to keep away from this locality:

> In the Atlantic States they appear not to understand that the rich silver mines of the Comstock are few in number. . . . The Comstock does not cover the whole State, nor does the bonanza extend through the whole Comstock. We now have miners enough and more than enough to work all of our mines, both paying and prospective. Let this be understood abroad, and let moneyless men give the Comstock a wide berth.[11]

But the *Enterprise* held out the advice to those willing to come to the West to enter the region's emerging sheep ranching: "Better cross the Sierras and seek employment on the sheep ranches of Southern California, than to stop here to starve while walking over the millions that lie buried deep in the rocky bosom of the big bonanza."[12]

From 1864 to 1878, fluctuations in the Comstock mining boom laid bare the uncertain future of the state's economy. When the Comstock failed after 1877, the structural weakness of the mining economy became evident. As an alternative to failing mining investments, many capitalists invested in stock ranching. In 1876, a Nevada sheep operator said: "Sheep are better than a government bond; you can tear off a coupon every six months half as big as the bond, and the bond is left as big as it was." Within a period of twenty years, after the Comstock's failure, stock raising operations assumed a major role in Nevada's economy with sheep starting to rival cattle as the preferred stock. Stock raising agriculture offered the promise of solid and stable economic development in Nevada. In 1881, in a travel article titled "Through Nevada's Desert," a correspondent for the *New York Times* wrote: "The people of Nevada, now that the mining future is so dubious, can see nothing ahead but agriculture, and, as it has been demonstrated that the soil is good, the great problem is water." Later, in 1885, the *Central Nevadan* of Battle Mountain noted: "There is no country in the world where sheep thrive so well, are so free from disease and attain such perfection as in Nevada."[13]

Limited Private Lands: The Law and Nevada's Ranges in the Late Nineteenth Century

In Nevada, the arid conditions determined economic development, especially for crop agriculture, as well as for grazing, and ultimately the failure of the state to attract a population base. Despite the small, sparse settlements around mining, and ranching towns along the railways, most of the public domain of Nevada remained unclaimed and unsurveyed. The unsettled land situation in Nevada challenged the federal land policy that encouraged the alienation and privatization of the public-domain lands.[14]

When Nevada became the thirty-sixth state in the Union on October 31, 1864, its statehood Enabling Act by Congress asserted ownership of the public domain by the United States and required the state to forever disclaim any right and title to the unappropriated federal lands. By this act, Congress granted 12,800 acres to Nevada for public buildings and education. From the late nineteenth century on, Nevada's federal lands have been a focus of debate regarding local versus federal authority, particularly in access and use of resources. In an arid land, water is one of the most crucial resources. The history of the practical operation of important land laws of the United States in relation to the range industry and its need for water is complex, but a brief consideration of these laws is important as they were critical in the larger development of the livestock industry in the region.[15]

In 1862, true to its promise in the election campaign of 1860, the Republican Party passed the Homestead Act a year after Congress organized the Territory of Nevada in March 1861. The Homestead Act offered heads of families free lands in the West to build homes and farms. Citizens who were heads of families (or who intended to become citizens) were entitled to claim 160 acres of unappropriated public domain. Improvements in terms of structures and land cultivation were assumed from the homesteader living on the land. The 1909 Enlarged Homestead Act offered 320 acres of nonirrigable lands suitable for dry farming as designated by the General Land Office of the Department of the Interior. The homestead law with its amendments operated with variable degrees of success and failure in the West. It was more effective in the humid lands of the Upper Mississippi Valleys. And while its success was moderate in portions of the Intermountain and Pacific regions, in Nevada it was largely inapplicable for the type of large-scale stock agriculture grazing that occurred on the state's marginal lands. Federal land policy failure in Nevada was legendary. In the period from 1864 to 1904, far fewer homestead entries were made in Nevada than in any other western public-domain state.

Other generous land acts similar to the Homestead Act aimed to promote settlement in the arid West: the Timber Culture Act, the Timber and Stone Act, the Desert Land Act of 1877, and the Carey Act of 1894. However, each of these failed to achieve settlement results. Some required the development of water supplies and irrigation. Under federal land alienation laws, complete land privatization failed. The result was the continued ownership of much of the West and almost all of Nevada by the federal government. What land was acquired resulted in the monopolization of water sources and rangelands by large livestock operations.[16]

In 1880 for Nevada, Congress responded with a two-million-acre land grant by which the state's citizens selected lands for purchase. Under the original school land grants to states dating from 1841 and earlier, sections 16 and 36 in every surveyed township were made to the states for the support of common schools. However, this system did not work in Nevada. First, Nevada was mostly unsurveyed, and secondly, such a survey would yield worthless lands far from water sources. School land grants under the old law were clearly not applicable for the support of Nevada's schools. The 1880 school land grant to Nevada permitted state authorities and citizens to select unappropriated nonmineral public lands within the state in not less than 40-acre parcels that proved to be desirable land tracts for privatization. The disposition of "selected lands" was to be made under regulations prescribed by the Nevada legislature.

The 1881 state legislative session initiated the school land-selling process of randomly chosen acreage in no less than 40-acre parcels. The lands selected were generally those with nearby water sources that cattle interests purchased in smallholdings. Control of riparian lands meant control of the range. Water monopolization by cattle interests took hold in this period and under the selected lands distribution and purchase process. Stock operators often strategically acquired ribbon-like parcels of land adjacent to streams and also around water sources or springs. The legislature set a price of $1.25 acre for the lands, requiring only 25 cents an acre as down payment and extending the terms of payments sometimes over a fifty-year time span. These generous credit terms caused a rapid purchase of school lands and, of course, a monopolization of water sources.[17]

Nevada's large stock firms, such as those owned by John Sparks, W. N.M. McGill or the Dangberg family, became major purchasers of state-selected school lands. Those large operations, as range scientist James A. Young observed, "deposited funds in Carson City to be drawn upon to meet the numerous annual credit payments for state school grant lands." These companies had a number of contracts for further land purchases. There was a close cooperation between the State Land Office and the large livestock operations. The monopolization in land ownership seemed a twofold purpose for the large firms: they owned the best lands along the streams for hay and winter-feed production and at the same time they excluded small operators and itinerants from watered ranges. Through the possession and control of water sources, then, the school land disposition system facilitated rangeland monopolization. At a lower level, small operations and livestock farms struggled to survive in Nevada. Their business needed to combine the prized hay pasture, irrigated lands, some good rangeland, and access to water on their home ranch of no more than 320 acres, as well as the indispensable use of the public domain. Small operators could only survive in the ranching business by producing hay and watering on private land for the winter season and grazing on the federal lands. For many this was beyond their limited resources.[18]

For the most part, public rangelands and mining districts remained unregulated from the federal level. Congress had no intention to impose control or close the open rangelands to settlement. It did insist by the late 1880s that the public range not be fenced by stock operators or graziers. However, in the absence of central government regulation and its uncertainty about western public rangelands and mineral lodes, local authorities had freely begun making decisions. For example, after the coming of homesteaders and sheep owners to Nevada in the late 1860s, cattle producers saw increased use of the rangelands. In 1873, the Nevada legislature passed a branding law in an attempt to establish order on the ranges. With lack of enforcement, however, the state's capacity to control the range problems failed. While Nevada began to enforce some range law, livestock operators often ignored it.[19]

Squatters settled on different corners of Nevada's public domain, but without assurance that in the future they could acquire such properties. Typically, those kinds of squatters declared their intent to file a preemptive claim under any of the land laws, although most of them did not prove up because they did not pay the standardized $1.25 per acre for the land their animals grazed. Yet, the land could not be taxed by the state because it was still public domain belonging to the federal government. But local authorities were interested in legislation to formalize squatting activities. In 1865, the legislature of Nevada enacted a law concerning property taxes, which permitted counties to assess what they considered "possessory property" on the public-domain lands or land claims upon which a squatter had not proved up. The assessment occurred on the improvement to the property, such as houses, outbuildings, barns, and livestock.[20]

Toward the end of the 1880s, after a rapid livestock population growth and especially the expansion of the sheep industry, lack of regulation and grazing controls exposed continuing quarrels over range resources. In general terms, the Nevada land situation was divided into three main categories: first, the greater part of Nevada was public domain, with a huge percentage of marginal lands remaining unclaimed; second, over half of the property privately owned belonged to the railroads and big ranch outfits; third, virtually all the remaining privately owned lands were situated around or close to water sources along streams adjacent to productive grasslands. The result was the bulk of Nevada private property was in large ranches monopolizing surface waters.[21]

In the 1880s, the state of Nevada saw an unprecedented expansion and growth of stock numbers on the ranges—both sheep and cattle. In June 1880, the *New York Times* assessed: "In the sage-brush lands, alfalfa, the cereals, and all vegetables flourish in profusion where water can be obtained, and Nevada is becoming an important stock-raising State." Because of a huge depression in the mining industry, many investors in the extraction of metals transferred their capital to the livestock business. These years also witnessed a rise in the number and weights of sheep slaughtered in Nevada.[22]

For the fiscal year ending June 30, 1885, Christopher C. Powning, U.S. surveyor general of Nevada, prepared a report on the state's extractive economic activities utilizing public lands. In those days, the sheep industry was, Powning wrote, "in a prosperous condition."

> There are 500,000 sheep on the ranges. Mutton sheep are valued at $2 per head, and lambs at $1,50. The average annual increase is 80 per cent. The wool clip averages 6 pounds, and good Nevada wool brings 16 cents per pound. There are no dry seasons here for sheep, and each year will witness a marked increase in this important, permanent and profitable industry.[23]

While economic winds favored sheep grazing, it lacked, however, support from the state authorities.

On range issues, the Nevada state government and cattle interests allied to the detriment of the sheep industry. Nevada's legislature actively sponsored legislation to undermine sheep grazing because of its migratory, itinerant, and "foreign" character—a reference to the fact that around this time non–Anglo Saxon sheepherders, especially Basque immigrants, were increasingly working as sheepherders. The state sought ways to limit competition for the range and assert the cattle industry's economic position. In February 1889, the Nevada legislature passed a stock trespass law and in 1893 the Nevada assembly enacted a new statute by declaring it unlawful "to herd or graze any live stock upon the lands of another without having first obtained the consent of the owner or owners of the land so to do." The state of Nevada tried to close access from the public lands to water streams, springs, or holes, which were surrounded by privately owned lands. The new trespass law made stockmen responsible for double the damages caused by their stock trespassing onto private property. These efforts, however, were unable to blunt the long-term growth of the sheep industry which would continue to increase in the coming decades.[24]

Use of public-domain lands in the intermountain states of Nevada and Utah was absolutely vital to the operations of big stock outfits. The grazing of the federal lands reduced significantly sheep production costs. The open-range system of livestock grazing offers a picturesque feature of western agricultural history of this period. Basically, the labor contracted to graze livestock by the companies—either the sheepherders or their counterparts in the cattle industry who rode horses, the cowboys—wandered at will over the federal lands looking for pastures that were marginal and remained in the public domain. For many years, livestock operators grazed stock freely on these lands. By century's end, however, every stock grower knew that the days of free and open range might be numbered. Although many livestock owners made fortunes and prospered under the conditions of open-range grazing, uncontrolled stock numbers on the range depleted the resources. Additionally, the competition among operators for forage brought conflict and damage to the ranges. As a result, early on, sheep grazing in these open ranges often faced opposition in the state, especially from the powerful cattle interests. After 1890, however, a number of developments offered opportunities for sheep on Nevada ranges.[25]

The Sheep Bonanza of the 1890s

As in the Great Plains, the early cattle industry in the Great Basin developed with open-range grazing. Low operating expenses with cheap labor, free forage, and open range made for large profits. In the late 1880s, severe droughts and cold winters occurred. The winter of 1889–90 was the disastrous "White Winter" on Nevada ranges, which wiped out the stock herds in the Great Basin. The losses during this winter surpassed all previous winters and wreaked havoc on northern Nevada ranges. Locally in the Great Basin the "White Winter" demanded a transformation of open-range grazing practices that now required the cultivation of hay crops to winter feed stock. In many cases, this meant smaller home ranch operations began to irrigate lands and even to plant alfalfa and large outfits stopped simply letting cattle run freely on the open range during cold and dangerous winters.[26]

After the "White Winter," as stock operators abandoned the classic open-range system, they placed greater emphasis upon the raising of hay for winter feeding. The process required greater capital investment, reducing profit margins. Stock operators of all kinds suffered great losses. Many were forced to quit their businesses. Others reduced their operations, and some transferred into sheep grazing. Bankruptcies were frequent. Generally, although cattle losses were more severe than sheep, numbers of stock on the rangelands in both cattle and sheep fell dramatically. In March 1890, C. S. Reynolds, a sheep broker from Nebraska, noted the following about the critical situation: "One reason for the heavy loss in Nevada is that no hay was raised during the last two years . . . The meagre supply of hay was given to the cattle, and the sheep had to rustle or starve."[27] The eleventh U.S. census of June 1, 1890, reported that the state of Nevada showed fewer sheep on farms (ranches) than many other western states, especially Montana, Wyoming, New Mexico, Arizona, Utah, Idaho, Oregon, California, and Colorado.[28]

The "White Winter" added to the hard economic times across Nevada as the mines were also in a terribly depressed condition. In August 1891, on a visit to San Francisco, Nevada governor Roswell K. Colcord said the state's mines did not prove attractive for further investment. Colcord declared that most mine locations were either exhausted or overpopulated for the work available. He warned people of California to stay put and not to go to Nevada. During the decade of 1890, a general mining depression, disastrous winters for livestock, and a general national depression after 1893 reduced the population of Nevada from 47,355 in 1890 to 42,335 by 1900 according to the National Census of 1900.[29]

The significantly reduced cattle numbers on the Nevada ranges that resulted from the winter of 1889–1890 would benefit sheep graziers. James Young writes: "The net immediate effect of the white winter was freedom for the range sheep industry to expand without competition from previously established cattle ranches." In addition to the severe winters, availability of range resources had also been altered by overgrazing, inappropriate agricultural practices, and droughts. The exhaustion of perennial grasses was followed by the invasion of annuals, shrubs, and big sagebrush—all of which had negative effects on cattle grazing, but none of which adversely affected sheep grazing. Consequently, the new economic situation after the "White Winter" opened up a promising period for the sheep industry of Nevada. In spite of the hardships the sheep industry had faced during the White Winter, the sheep industry would go on to make advancements that frustrated the cattle community in Nevada.[30]

After the "White Winter," as the Nevada ranges became a major producer of sheep, sheep breeders stocked Nevada ranges with hundreds of thousands of animals. In the years 1890–1899, according to the agricultural census of 1900, the total sheep population in Nevada increased from 273,469 to 568,251 head, more than the total number of neat cattle (see table 1.1). Dispersed over wide range areas, market centers grew up with packinghouses in Denver and Omaha that connected the West with the main national markets.[31]

Meanwhile, the Nevada state legislature continued discussing the competition for range sources, frequently advocating for cattle interests. While the state assembly proposed several strategies to exclude sheep from state ranges, little was accomplished. In February 1895, John H. Weiland from Elko filed a bill in the Nevada legislature to require licenses

Table 1.1. Total number of sheep and neat cattle in the state of Nevada, 1860–1900

Nevada	1860	1870	1880	1890	1900
Total number of sheep	376	11,018	280,695	273,469	568,251
Total number of neat cattle	5,471	31,516	216,823	210,900	304,131

Source: U.S. Census Office, Census Reports. Vol. V, "Twelfth Census of the United States, Taken in the Year 1900," Agriculture. Pt. 1, "Farms, Live Stock, and Animal Products" (Washington, D.C.: Government Printing Office, 1902), 704, 708.

for sheep grazing in Nevada. The bill, Bill N. 90, was intended to limit itinerant sheep operations, particularly in consideration of the fact that oftentimes out-of-state companies from Utah, California, or Idaho sent large herds of sheep into Nevada, who employed immigrant labor such as the Basques. This bill proposed grazing licenses on a preference basis according to the total of sheep unit. Most importantly, the bill proposed licenses that could be as much as $250 annually on large sheep herds. It provided, furthermore, "that nothing in this Act contained, shall be construed as to require the procurement of more than one license for the same sheep, in the same county, for the same year."[32]

The bill responded to the perceived threat of the still-expanding sheep industry and made it a misdemeanor for any person to graze sheep without the proper license. Under this bill, county sheriffs were the sheep license collectors and in charge of inquiries and detailed examinations. But when the bill came up to vote, the legislature was deeply divided. Ultimately, the sheep license bill lost by a vote of eighteen to nine. Nevertheless, the session established a special "Sheep Law Committee" consisting of three members from the assembly and two from the senate to formulate a measure that would satisfy cattle and sheep interests. The members of the committee consisted of two cattlemen and two sheepmen, as well as one disinterested party. As Patrick L. Flanigan, Republican representative from Washoe County, put it: "This is a question of very great importance, and as we fully realize the necessity of a bill protecting both cattle and sheep interests."[33] But these legislative efforts remained controversial. As the *Nevada State Journal* declared on February 21, 1895, by attempting to pass the sheep license bill, the state legislature was working "at the quickest speed possible to kill the sheep industry in Nevada." The article remarked that the sheep industry should be considered one of the main economic sectors in Nevada. It argued that the economic development in Nevada was largely based on livestock ranching—either sheep or cattle. Given the arid climate and poor lands for crop agriculture, it explained how sheep ranching naturally grew and prospered in Nevada. The *State Journal* contended that the development of the sheep industry held the key to the state's future: "There are many persons engaged in it, and considerable [amounts] of the revenue for the support of State and county governments is derived from it. It utilizes pasturage in Nevada which other wise would be wasted. It gives employment to a good many persons, who, if the business is crippled by unfriendly legislation, would have to leave the State."[34]

On March 11, 1895, the Sheep Law Committee proposed in the senate a substitute act for Assembly Bill N. 90, establishing further regulation of the open-range sheep industry in Nevada for the local government. The state substitute bill was then approved by the majority and approved by the governor of Nevada, John E. Jones.[35]

The *Reno Weekly Gazette and Stockman* spoke for many Nevada ranchers and landowners when it urged a solution to bring order to the rangelands, regardless of the competition between the sheep and cattle interests. The article urged that all landowning graziers, be they either sheep or cattle people, be interested in suppressing the presence of "roving bands of stock":

> Decent stockmen in Nevada have been at the mercy of roving bands of stock that were owned by aliens and run where they pleased, taking the best of everybody's feed without paying rent, taxes or anything else. . . . This would simply have had the effect to kill the business, and the State and counties would have lost an important industry. Nevada has so few lines that can be made productive that every one who gave the matter any intelligent consideration felt that it would be very bad to kill off the sheep industry.[36]

The majority of resident stock operators (sheep and cattle people) who owned land, either large or small, saw transient sheep grazing on the public domain as a threat to established range practices and therefore a threat to law and order. This rural public sentiment against the itinerant sheep grazing on federal lands was in accord with the yeoman farmer private property tenets of Jeffersonian Democracy that the privatization of the majority of the small farms in the public domain made possible.[37] In Nevada, however, as mentioned earlier, the prevailing land laws did not lead to the privatization of most of the public domain as had occurred in other states. Assemblyman Patrick Flanigan, a resident sheep grazier, backed the sheep license act because it undermined the position of the nonresident stock growers, especially Basque and Portuguese immigrant herders and the Mormon sheepmen of Utah. Their herds came in from out of state and used the Nevada ranges while the owners contributed nothing in the way of tax revenue to the state. According to Flanigan, nine-tenths of the state's woolgrowers community strongly supported the bill. He said that both cattlemen and sheepmen were satisfied with the sheep license bill. Part of the law was a reaction against the recent European immigrants and the out-of-state companies who employed herders to bring sheep into the state. In particular, an anti-Basque sentiment was explicit in Flanigan's statements.[38]

In 1897, the legislature revised the law to compel all the sheep owners to pay a license to the counties in which they operated, unless they owned one acre of land for every two head of sheep on the range. Although the new law was approved and celebrated by many established stock operators, its application was problematic. The itinerant graziers who moved sheep from one section of grazing lands to another were usually difficult to follow as they moved across county lines. Additionally, some graziers who did not own land found an alternative way to avoid paying for a license by leasing parcels. In this case, legally, leasing or owning land was the same. Since the railroads were the largest landowners, their land agents saw the new regulation as another way to make money through leasing railroad lands. Sheep raisers began to lease railroad land because it was much cheaper than the amount they had to pay under the new state sheep tax laws. Consequently, the railroad companies leased a large number of poor and worthless lands to sheep operators.[39]

On January 26, 1897, a taxpayer from Golconda wrote a letter to the editor of the Salt Lake City *Daily Tribune* complaining about these irregular activities:

> The result is that except to help the railroad company lease worthless land the law is a blank failure. The only parties satisfied with it are the railroad companies and the tramp sheep-owners. The sheep-owner rents a few townships (that is the odd sections) of alkali land at, say, $10 per township—which likely he

> never sees at all. Certainly he never occupies—and with his lease in his pocket claims the right to range over the whole State. The question may be settled to the satisfaction of the courts, the railroad company and the tramp sheepmen, but scarcely to the satisfaction of the taxpayers and legitimate stock-raisers.[40]

Clearly, this Nevadan described fraud and a failure of the law. The law did not satisfy those from the livestock community who wished for a restriction on those whom they saw as outsiders creating chaos on the Nevada ranges. Remedies to the situation seemed almost impossible.

In the decade from 1890 to 1900, uncertainty and confusion among stock operators increased. They feared that another "White Winter" might occur. Although uncertainty for the future worked against investment in cattle ranching, sheep operations seemed more secure. The sheep population actually increased as state efforts failed to exclude sheep outfits. Cattle operators were frustrated to observe sheep bands efficiently grazing even on deteriorated ranges. In response, many cattlemen converted to sheep grazing because at that time it was the most profitable business. The future of the cattle industry seemed full of difficulties.[41]

Wool Tariffs and the Nevada Range

In the late nineteenth century, American agriculture witnessed an unprecedented agricultural expansion at all levels, a process that was stimulated by national industrialization. Furthermore, improvements in transportation and marketing facilities encouraged agricultural regional development. This explains also the growth of the western sheep ranching in those decades. In addition, the international markets gave a prominent impetus to Nevada's sheep industry. During the last three decades of the nineteenth century, the global economy saw a prominent increase in the circulation of American manufacturing and agricultural products. In the post–Civil War period, unlike England, the United States was endowed with abundant land resources for extensive livestock farming in the Far West. In the matter of exports, American wool expanded to the markets abroad to supply manufacturers in Europe.[42]

In the late nineteenth century, the United States went back and forth between increasing and limiting protectionist policy in response to national interests. The American sheep industry also had a stake in protectionism. With the expansion of the sheep industry in the West, prices for its products were affected by congressional policies on wool tariffs. From the late 1880s, the tariff question became a controversial subject between the western woolgrowers and Congress that persisted well into the twentieth century. "The wool and woolens schedule," tariff historian Frank W. Taussig wrote, "had become the most important and most sharply debated part of the tariff system." Changes and adjustments in tariff rates of wool and sheep products affected their market prices and determined investments in the industry and ultimately the number of stock on the ranges.[43]

Generally, from the 1870s to the early twentieth century, American exports exceeded imports. The protective tariff policies helped build many American industries and position them for expansion into international markets, and the sheep industry was no exception. Directly, tariff policies had effects on various economic sectors (such as the sheep industry's wool production) and, thereby, the tariff was unsurprisingly a constant issue of post–Civil War American politics and the rivalries between the Republican and Democratic Parties. The United States had adopted strong protective policies during and after the American Civil War, policies that aimed to encourage domestic production and meet foreign competition, especially against the United Kingdom and Germany. It was in 1883 when, for the first time after the Civil War, the tariff legislation was revised and a new law was enacted accomplishing a reduction of revenues. This tendency endured until 1890 when, unexpectedly, Congress passed a new tariff act—often referred to as the McKinley Bill—which strengthened the protective system. Responding to its industrial lobby, the Republican Party supported this protective tariff. Economic historian Sidney Ratner noted that the protection of various national industries against foreign competition effectively excluded the flow of imports from Western Europe.[44]

In the late 1880s, before the new tariff act was passed in 1890, the tariff question was a major issue of public debate and, of course, in the political arena. The two major parties became polarized on the tariff issue. If the Democrats had embraced the idea of free commerce and attacked protection (despite the divergent opinions on this issue in the party), the Republicans enthusiastically adopted a position that insisted upon the enlargement of duties. The 1890 act reestablished duties on wool enforced before 1883, all of which favored national woolgrowers and manufacturers.[45]

In 1893, when Democratic candidate Grover Cleveland victoriously returned to the White House, the political debate resumed upon the tariff issue. Cleveland, who opposed strongly protective legislation, received the full support from tariff reformers who asked for a change in international trade issues. One of the first tasks of this administration was to revise the tariff legislation downward, asserting the traditional Democratic Party's position favoring free trade or the liberty of commerce. In the fall of that year, the Committee of Ways and Means began to draft a tariff bill. Under the chairmanship of William L. Wilson, Democrat of West Virginia, the house committee was composed of seventeen members: eleven Democrats and six Republicans. Wilson drafted the tariff bill by proposing the reduction of duties on manufactured goods, including the products of wool and sugar on the free list.[46]

At the time the Congress was discussing a reduction of tariff duties, an anti-tariff reform lobby also formed. On November 30, 1893, the Nevada Woolgrowers Association wrote a letter to the Representative Julius C. Burrows, one of the Republican members of the Committee of Ways and Means, asking him to oppose the Cleveland administration's proposal to liberalize tariff rates on wool. The letter is a clear advocate of the sheep industry in Nevada:

> With free wool we could not have one cent profit per fleece. We ship in the grease and have a shipment of 70 pounds of dirt for every 30 pounds of wool. If we could figure on the price at New York being 30 cents per pound for washed wool (as it would be under free wool) it would leave us 13 cents per head for expenses of running our sheep. That is, if we could bring the sheep to the shearing pen for 13 cents, it would still cost us 17 cents per pound to get the wool clipped, washed and freighted to New York. You can easily see that we cannot depend upon a profit under any such an arrangement.[47]

Nevada woolgrowers defended sheep ranching as "a regular step in State growth." To them, the increase in the industry benefited every branch of trade locally. Fearful sheepmen claimed to Burrows that the value of every sheep was cut off because of tariff agitation and if the bill was finally approved, it would cause woolgrowers huge production losses and result in lower American wool prices. They also cited statistics to show that the intermountain states could be more prosperous economically with protective legislation. Nevada woolgrowers contended bitterly, "Common sheep and common wool will grow anywhere the finer breeds can be raised, and the only reason the common wool has not come in is because of the tariff."[48] As in Nevada, ranchers from other western states complained to their representatives about the Democratic tariff reform bill.

Nevertheless, in 1894, Congress, following the guidelines proposed by the Committee of Ways and Means, passed the tariff reform bill. One of the main provisions of the act was the removal of the duty on wool, something that would lead to real hardship for American woolgrowers. Continuing the economic and political jockeying, however, several years later, with the passage of the tariff act of 1897, Republicans would again proceed to raise tariff barriers, strengthening duties on raw wool by raising the *ad valorem* tariff rate to 50 percent. With strong economic interests concerned in the western ranching states, political forces had pushed Congress to place high duties on wool imported in the United States. In the eyes of many westerners, a free-trade policy undermined the livestock economy of the West.

Despite the benefits to American woolgrowers, the rewards of the 1897 law were not necessarily evenly distributed. In December 1897, after the passage of the tariff act, Taussig wrote: "While there is a duty on wool, intended for the benefit of the farmer, that duty will almost certainly inure chiefly to the advantage of the ranchmen of the more distant western states." This is to say, the large-scale production practiced in the grazing of vast herds on the public ranges of the West obtained larger profits than any small farmer could achieve on their various farms as any slight or even dramatic increase to the price of sheep products occasioned by a higher tariff rewarded significantly large-scale producers. Industry-wide, however, it was clear the sheep industry was flourishing in Nevada. In the census of 1900, the state of Nevada reported the production of 4,842,500 pounds of wool compared to 1,450,868 pounds in the 1890 census.[49]

Summary

The sheep boom that Nevada witnessed in the last two decades of the nineteenth century discussed in this chapter was similar to that experienced in other western public-land states. In the case of Nevada, this boom laid the foundation for the growth in the Basque sheepherding industry in that state during those decades and in the twentieth century.

Demand for meat and wool in the United States and abroad spurred the expansion of Nevada's sheep industry. In comparison with cattle ranching, the open-range sheep industry had higher returns on investment over the short term. After the "White Winter" of 1889–1890, the economic profitability of sheep grazing in the Great Basin encouraged old stock owners and newcomers to enter this industry. Even after this devastating weather event, the sheep population would increase considerably in Nevada.

The economic success of sheep outfits in Nevada depended largely upon the free use of public-domain lands, secure winter feed, water resources, and employment of low wage, dependable agricultural labor. By the turn of the century, the state of Nevada still offered the ideal setting to produce sheep at the lowest cost possible because of the still abundant unclaimed public domain and its grazing lands. The lack of any regulation of grazing on these lands in terms of numbers and season of graze and access to cheap labor to oversee great herds of sheep presented tremendous opportunities for profit in the sheep industry. Sheep owners demanded a pliable labor force willing to work for long periods of time herding large flocks of sheep, for little pay, and often in isolation under difficult conditions in the harsh environment of the Great Basin mountainous desert. In the late nineteenth century, immigrants from southern and eastern Europe were often the ones who found themselves in these jobs.

Many Basque immigrants entered low-paying sheepherding work in Nevada, drawn to the state by already-established Basque ranching operations. In the 1870s, a handful of Basque immigrants, who had come to California during the gold rush, left that state to take up livestock opportunities in northeastern Nevada. By the late 1880s, these early Basque arrivals in Nevada had built up prominent stock ranches. They found willing and economical labor in their homeland among their family members and neighbors, many of whom would emigrate to work in these operations. Among the Basques, it became a common practice to make an effort to employ their own ethnic group on their ranches instead of hiring local ranch hands already present in the Great Basin. These Basque pioneer ranchers forged a chain of migration that lasted for years.

The next chapter brings the emergence of Basque immigration to Nevada into view, and reveals the important ways that the Basque immigrants became crucial members of that state's industrial growth in the sheepherding industry.

Notes

1. This region of the Far West is an interior drainage basin consisting of a diversified topography primarily composed of mountain ranges, highland valleys, steep gradient mountains, deep canyons, and prehistoric lakes. This semi-arid region extends from the present eastern half of Utah, small areas in southwestern Wyoming and southeastern Idaho, some parts of the east-central Sierra Nevada in California, southeastern Oregon, and most of the state of Nevada. The main focus of the present study is on Nevada's Great Basin. Samuel G. Houghton, *A Trace of Desert Waters: The Great Basin Story* (Salt Lake City, Utah: Howe Brothers, 1986), 23–25; for treatment of the physical environment of the Great Basin see also: John McPhee, *Basin and Range* (New York: Farrar, Strauss, and Giroux, 1981); Bill Fiero, *Geology of the Great Basin* (Reno: University of Nevada Press, 1986).
2. Effie M. Mack, *Nevada: A History of the State from the Earliest Times through the Civil War* (Glendale, Calif.: The Arthur H. Clark Company, 1936), 19–38.
3. "Wealth in Wool," *Reno Weekly Gazette and Stockman*, April 18, 1889, 8.
4. Russell R. Elliott, with the assistance of William D. Rowley, *History of Nevada* (Lincoln: University of Nebraska Press, 1973), 115–22.
5. Elliott, *History of Nevada*, 115; Clel Georgetta, "Sheep in Nevada," *Nevada Historical Society Quarterly* 8, no. 2 (Summer 1965): 18.
6. Edward N. Wentworth, *America's Sheep Trails: History, Personalities* (Ames: Iowa State College Press, 1948), 165–69; Byrd W. Sawyer, *Nevada Nomads: A Story of the Sheep Industry* (San Jose, Calif.: Harlan-Young Press, 1971), 1–13; Clel Georgetta, *Golden Fleece in Nevada* (Reno, Nev.: Venture Publishing, 1972), 6–26.
7. Elliott, *History of Nevada*, 116–18.
8. Mack, *Nevada*, 200–08, 437–55; James W. Hulse, *The Silver State: Nevada's Heritage Reinterpreted* (Reno: University of Nevada Press, 1991), 74–79, 133–35.
9. Elliott, *History of Nevada*, 118–21; Sam P. Davis, *The History of Nevada:* Volume I (Reno, Nev.: Elms Publishing Company, 1913), 640–45; Edna B. Patterson, Louise A. Ulph, and Victor Goodwin, *Nevada's Northeast Frontier* (Sparks, Nev.: Western Printing and Publishing Company, 1969), 211–13; Hubert H. Bancroft, *History of Nevada, Colorado, and Wyoming, 1540–1888*, vol. 25 of *The Works of Hubert Howe Bancroft* (San Francisco, Calif.: History Company, 1890), 243.
10. James A. Young and B. Abbott Sparks, *Cattle in the Cold Desert* (Logan: Utah State University Press, 1985), 48–55; from the *Humboldt Register* as noted in *Territorial Enterprise*, August 6, 1872, 2; Sawyer, *Nevada Nomads*, 15, 17; a contemporary sheepherding manual read: "In the management of sheep, how to procure the most profit from the flock is the greatest consideration. It is not exactly how to increase the flock most rapidly, nor to produce the heaviest carcasses or fleeces, but to produce such animals as will return the most money for the expenditure and labor involved." Henry Stewart, *The Shepherd's Manual: A Practical Treatise on the Sheep: Designed Especially for American Shepherds* (New York: O. Judd Co., 1876), 33.
11. An article from the *Territorial Enterprise* as reprinted in the *New York Times* discouraged easterners from going West: "No Miners Needed in Nevada," *New York Times*, March 21, 1875, 2.

12. Ibid.
13. Elliott, *History of Nevada*, 171–73; *Emmetsburg Palo Alto Pilot*, March 23, 1876; "Through Nevada's Desert: A Country with Many Possibilities of Development," *New York Times*, March 20, 1881, 5; as reprinted in *Nevada State Journal*, August 18, 1885, 3.
14. Davis, *The History of Nevada*, 640; Mack, *Nevada*, 29–31; Hugh A. Shamberger, "Evolution of Nevada's Water Laws, as Related to the Development and Evaluation of the State's Water Resources from 1866 to about 1960," *Water-Resources Bulletin* 46 (Carson City: Nevada Division of Water Resources, 1991), 3–6; Donald J. Pisani, "Federal Reclamation and Water Rights in Nevada," *Agricultural History* 51, no. 3 (July 1977): 540–41.
15. Mack, *Nevada*, 266–67; Elliott, *History of Nevada*, 84–85; *Biennial Report of the Surveyor-General and State Land Register of the State of Nevada, 1909–1910* (Carson City: State Printing Office, 1911), 28–30; Eleanor Bushnell, *The Nevada Constitution: Origin and Growth* (Reno: University of Nevada Press, 1965), 117–18; Michael J. Brodhead, "Accepting the Verdict: National Supremacy as Expressed in State Constitutions, 1861–1912," *Nevada Historical Society Quarterly* 13, no. 2 (Summer 1970): 9–10; Michael J. Brodhead and James W. Hulse, "Paul W. Gates, Western Land Policy and the Equal Footing Doctrine," *Nevada Historical Society Quarterly* 29, no. 4 (Winter 1986): 234–36, 239.
16. Smith, *Virgin Land*, 195–96; Eric Foner, *Free Soil, Free Labor, Free Men: The Ideology of the Republican Party before the Civil War* (New York: Oxford University Press, 1970), 31–38, 175; Paul W. Gates and Robert W. Swenson, *History of Public Land Law Development* (Washington, D.C.: Government Printing Office for Public Land Law Review Commission, 1968), 393–414, 638–54; Young and Sparks, *Cattle in the Cold Desert*, 89–98; Mack, *Nevada*, 415, 420; *Biennial Report of the Surveyor-General and State Land Register of the State of Nevada, 1909–1910*, 9–21; Donald J. Pisani, *To Reclaim a Divided West: Water, Law, and Public Policy, 1848–1902* (Albuquerque: University of New Mexico Press, 1992), 186–88.
17. Mack, *Nevada*, 420–22; *Appendix to the Journals of Senate and Assembly of the Tenth Session of the Legislature of the State of Nevada* (Carson City: State Printing Office, 1881), 10 [Appendix]; *Report of the Surveyor General and State Land Register of the State of Nevada, for the Years 1881 and 1882* (Carson City: State Printing Office, 1883), 21–22; *U.S. Treasury Department, Bureau of Statistics, Report in Regard to the Range and Ranch Cattle Business of the United States*, by Joseph Nimmo Jr., (Washington, D.C.: Government Printing Office, 1885), 430–44; Romanzo Adams, "Public Range Lands: A New Policy Needed," *American Journal of Sociology* 22, no. 3 (November 1916): 324–51; Bancroft, *History of Nevada*, 315–17; Young and Sparks, *Cattle in the Cold Desert*, 94–98; John M. Townley, "Management of Nevada's State Lands, 1864–1900," *Journal of the West* 17, no. 1 (January 1978): 63–73.
18. Young and Sparks, *Cattle in the Cold Desert*, 96, 97–98; Marion Clawson, "Range Lands of Northeastern Nevada: Their Proper and Profitable Use (A Progress Report)," Bureau of Agricultural Economics, USDA (Washington, D.C.:

Government Printing Office, 1938), 4–6; *Statutes of the State of Nevada Passed at the Fourteenth Session of the Legislature, 1889* (Carson City: State Printing Office, 1889), 129; Townley, "Management of Nevada's State Lands," 63–73; Adams, "Public Range Lands," 324–51.

19. Clawson, "Range Lands of Northeastern Nevada," 13–14, 16–17; *Statutes of the State of Nevada Passed at the Sixth Session of the Legislature, 1873* (Carson City: State Printer, 1873), 99–101; Halina M. Stewart, "The Nevada State Detectives," *Nevada Historical Society Quarterly* 49, no. 1 (Spring 2006): 25–28; see also, Velma Stevens Truett, *On the Hoof in Nevada* (Los Angeles, Calif.: Lorrin L. Morrison, 1950).
20. Young and Sparks, *Cattle in the Cold Desert*, 97–98; *Statutes of the State of Nevada, Passed at the First Session of the Legislature, 1864–5* (Carson City: State Printer, 1865), 343–44; Romanzo Adams, *Taxation in Nevada: A History* (Reno: Nevada Historical Society, 1918), 100–110.
21. Elmer O. Wooton, *The Public Domain of Nevada and Factors Affecting Its Use*, USDA Tech. Bulletin no. 301 (Washington, D.C.: Government Printing Office, Apr. 1932), 37–49; Adams, "Public Range Lands," 331–35, 341–46.
22. "Farms in the Great West: Opportunities for Settlers upon the Public Lands," *New York Times*, June 4, 1880, 2; Elliott, *History of Nevada*, 171–74; Patterson, et al., *Nevada's Northeast Frontier*, 269–71, 279; Stephen Powers, *The American Merino: For Wool and for Mutton* (New York: Orange Judd Company, 1907), 276.
23. Christopher C. Powning, "Something about Public Lands, Mining, Grazing, Sheep Husbandry, and Other Nevada Matters," *Report of the U.S. Surveyor General of Nevada for the Fiscal Year Ending, June 30th, 1885*.
24. William D. Rowley, *Reclaiming the Arid West: The Career of Francis G. Newlands* (Bloomington: Indiana University Press, 1996), 50–52; *Statutes of the State of Nevada, Passed at the Sixteenth Session of the Legislature, 1893* (Carson City: State Printing Office, 1893), 30–31.
25. William Rowley, "From Open Range to Closed Range on the Public Lands," in *Land in the American West: Private Claims and the Common Good*, eds. William G. Robbins and James C. Foster (Seattle and London: University of Washington Press, 2000), 96–98; Louise E. Peffer, *The Closing of the Public Domain: Disposal and Reservation Policies, 1900–1950* (Stanford, Calif.: Stanford University Press, 1951), 26; Young and Sparks, *Cattle in the Cold Desert*, 89–100.
26. Walter Prescott Webb, *The Great Plains* (Lincoln: University of Nebraska Press, 1981), 226–230; Edna Patterson, "Early Cattle in Elko County," *Nevada Historical Society Quarterly* 8, no. 2 (Summer 1965): 6–12; James A. Young, "Hay Making: The Mechanical Revolution on the Western Range," *Western Historical Quarterly* 14, no. 3 (July 1983): 311–13.
27. "The Sheep Interest," *Daily Nevada State Journal*, March 14, 1890.
28. Census 1890, total number of sheep and wool on farms by states in the Western Division: Montana, 1,859,016; Wyoming, 712,520; Colorado, 717,990; New Mexico, 1,248,970; Arizona, 102,427; Utah, 1,014,176; Nevada, 273,469; Idaho, 357,712; Washington, 265,267; Oregon, 1,780,312; California, 2,475,140. 52d

Cong. 1st Sess. H.R., Mis. Doc 340, Part 20, Reports of the Statistics of Agriculture in the United States, Agriculture by Irrigation in the Western Part of the United States, and Statistics of Fisheries in the United States at the Eleventh Census: 1890 in *The Miscellaneous Documents of the House of Representatives for the First Session of the Fifty-Second Congress, 1891–92, vol. 50, part 10* (Washington, D.C.: Government Printing Office, 1896), 236.

29. Elliott, *History of Nevada*, 171, Table 1; "Don't Rush to Nevada," *New York Times*, August 30, 1891, 1; *Twelfth Census of the United States, Taken in the Year 1900, William R. Merriam, Dir. Population, Part 2* (Washington, D.C.: United States Census Office, 1902), 632.
30. Young and Sparks, *Cattle in the Cold Desert*, 135, 136.
31. Lane, "The Cultural Ecology of Sheep Nomadism," 56–59; Richard White, *"It's Your Misfortune and None of My Own": A History of the American West* (Norman: University of Oklahoma Press, 1991), 222–27; Powers, *The American Merino*, 209–51; On September 27, 1893, the Salt Lake City *Daily Tribune* reported about some "heavy sheep shipments from Nevada" that ran directly to the main markets in San Francisco and to the midwestern and eastern cities of Chicago and New York. *Daily Tribune*, September 27, 1893, 7.
32. *The Journal of the Assembly of the Seventeenth Session of the Legislature of the State of Nevada, 1895* (Carson City: State Printing Office, 1895), 103–4; *Reno Weekly Gazette and Stockman*, March 21, 1895.
33. *Reno Weekly Gazette and Stockman*, March 21, 1895; *The Journal of the Assembly of the Seventeenth Session of the Legislature of the State of Nevada, 1895*, 180.
34. "To Kill the Sheep Industry," *Nevada State Journal*, February 21, 1895.
35. *The Journal of the Assembly of the Seventeenth Session of the Legislature of the State of Nevada, 1895*, 188, 207–8.
36. *Reno Weekly Gazette and Stockman*, March 21, 1895.
37. William D. Rowley, *U.S. Forest Service Grazing and Rangelands: A History* (College Station: Texas A&M University Press, 1985), 44; Phillip O. Foss, *Politics and Grass* (Seattle: University of Washington Press, 1960), 12–13.
38. "The Sheep License Act," *Nevada State Journal*, May 5, 1895.
39. *The Journal of the Assembly of the Eighteenth Session of the Legislature of the State of Nevada, 1897* (Carson City: State Printing Office, 1897), 182–83, 198, 219; *The Journal of the Senate of the Eighteenth Session of the Legislature of the State of Nevada, 1897* (Carson City: State Printing Office, 1897), 203.
40. *Daily Tribune*, January 28, 1897, 7.
41. Young and Sparks, *Cattle in the Cold Desert*, 219–35.
42. Sidney Ratner, James H. Soltow, and Richard Sylla, *The Evolution of the American Economy: Growth, Welfare, and Decision Making* (New York: Basic Books, 1979), 255–73, 387–89; Richard White has assessed that the economic potential in the American West was firmly intermingled with worldwide markets: "The West possessed an extractive economy that depended on outside markets, outside capital, and, most often, skills and technologies imported from the outside. Westerners had learned to look anxiously to eastern and European markets and investors.

Outsiders controlled their economic fate." White, *"It's Your Misfortune and None of My Own,"* 267, 268.

43. The history of domestic sheep industry is largely a continuous hard bargaining between tariff rates and the economic health and development of the global market. The 1890s is considered a decade of high tariffs that inhibited American foreign trade, with the exception of the year 1894. Frank Taussig wrote the passage of the 1890 Tariff Act—the so-called McKinley Tariff Act—marked "a new phase in our [American] tariff history and in the protective controversy." Frank W. Taussig, *The Tariff History of the United States* (New York: Putnam, 1910), 256, 251; Frank W. Taussig, "The Tariff Act of 1897," *Quarterly Journal of Economics* 12, no. 1 (October 1897): 42–69.
44. Ratner, et al., *Evolution of the American Economy*, 387–9; Taussig, *Tariff History of the United States*, 230–51.
45. Taussig, *Tariff History of the United States*, 255–66.
46. Arthur B. Darling, "Grover Cleveland and National Finances," *American Scholar* 3, no. 2 (Spring 1934): 144–54; Roy G. Blakey and Gladys C. Blakey, *The Federal Income Tax* (New York: Longmans, Green, 1940), 12–13.
47. *Reno Evening Gazette*, December 19, 1893.
48. Ibid.
49. Frank W. Taussig, "The United States Tariff Act of 1897," *The Economic Journal* 7, no. 28 (December 1897): 595, 592-98; *Twelfth Census of the United States, Taken in the Year 1900, Agriculture, Part 1, Farms, Live Stock, and Animal Products* (Washington, D.C.: United States Census Office, 1902), 679; *Miscellaneous Documents of the House of Representatives for the First Session of the Fifty-Second Congress*. 236.

2

Becoming Herders

Basque Immigration, Labor, and Settlement in Nevada, 1880–1910

Born April 1, 1892 in Markina, Biscay, Pedro Barinaga would be later known in Nevada as "Buckaroo Pete." He grew up on the family's farmstead in Biscay. His father worked in a rock quarry located a few miles from their home, although like most working men in the vicinity, his father spent very little time at home. One gray day, the young Pedro was repairing the homestead's fence with his cousin who was about to emigrate and join his own father, Pedro's uncle, in the United States where he was working as a sheepherder in Nevada. Pedro's cousin asked him if he wanted to join him in this adventure. Pedro had previously heard many stories from Basques in America, especially from reading the letters from his uncle who was living in Battle Mountain, Nevada, about the work opportunities and the high rate of pay there. In the end, after his cousin's urging, Pedro was convinced to go and he decided to emigrate. Pedro's father paid his ticket.[1]

In early March 1911, Pedro Barinaga departed by train to the port of Le Havre in northern France where he boarded the ship *La Gascogne* for America. He was very nervous and sad to be leaving his family and homeland, but he was not alone in this trip. With him on the ship were at least half a dozen other men from the same town of Markina leaving the Old Country to become sheepherders in the West. It was their first time on a ship and away from home. Furthermore, on the same steamship, other fellow countrymen from other parts of the Basque Country traveled with the same purpose. They included individuals from several towns around the Basque Country, including Deba, Mutriku, and Zestoa.[2]

On March 14, Pedro and the other Basques arrived in Ellis Island in New York, the major port of entry for Basque immigration. Once in the United States, an immigrant agent and businessman, Valentin Aguirre, received the group of Basques and conducted them to his boardinghouse, *Casa Vizcaína* (Biscayan House), in New York City. Aguirre provided all kinds of assistance, advice, and services to the new immigrants, facilitating their transitory stay in New York before they proceeded across the country to the West.[3] In Penn Station, Pedro—now "Peter"—Barinaga embarked on a train for St. Louis, Missouri, where he took another train to reach his last destination: Battle Mountain, Nevada.

After the long railway journey, Barinaga jumped off the train at the station in Battle Mountain. Following the instructions his uncle had provided, he looked for the Basque boarding house owned by Juan Belaustegui, another Basque immigrant, who had arrived in Nevada in 1902. The next day, Pete Olabarria, the Basque foreman of the ranch where Barinaga intended to work, brought Pedro to Esteban Mendibe's workwear store in town, where he was furnished with the clothing and tools necessary to work on the ranch. The employer paid all the charges for Pedro, who only had three dollars in his pocket. Later these charges would be deducted from his salary. Very soon after his arrival, it was time for sheep shearing and lambing. Lots of ranch work was at hand. Barinaga had never before worked with sheep, but he was willing to learn and ready to do the tasks.

After the lambing season, which lasted approximately one and a half months, Pete found himself without employment. Despite the stories that he had heard before coming about the high pay, when the short lambing season finished, he found himself without any money. Indeed, after his work, he owed money to his employer. The bills he'd had to pay for the clothing and other supplies, as well as the lodging costs at the Belaustegui's rooming house, had used it all up. Eventually, he found work with a German rancher. For the next three years, from 1911 to 1914, Pete worked from sunrise to sunset as a ranch hand. After working through the three-year contract, Pete's employer only gave him a $90 check. At that time, the average pay for an employee like Pete was one dollar a day paid every three months. But communication and language barriers enabled the German rancher to exploit Pete. Once again, the pay was so low that he had not been able to save any money. He decided to find another ranch on which to work. He went to work for Marcos

Rekalde, a small Basque rancher in Battle Mountain. In the ensuing years, Pete Barinaga would go on to work as a ranch hand for various livestock operations in the area of Battle Mountain where he would go on to obtain a good reputation as a stockman and trapper.[4]

The expansion of sheep agriculture in the 1890s resulted in the employment of significant numbers of workers, largely European immigrants—including those from the Basque Country like Pete Barinaga and his cousin—to labor in the most desolate rangelands of the Great Basin. During the 1890s, there was demand for contract labor from sheep operations, some of which were owned by Basques. Early Basque pioneer stock operators had begun recruiting Basque workers from the Old World, and the efforts of these early contractors set in place the immigration pattern between the Basque Country and the American West.[5]

Historian David Igler has introduced the concept of "Industrial Cowboys" to refer to those working as a part of commercial livestock ranching enterprises organized around land investments, profit-making, and food production, such as those occurring in Nevada at this time. The industry "thrived by engineering natural landscapes and mobilizing large labor forces." The concept contradicts the traditional mythic symbol of the free-spirited cowboy ranch worker always on the move to the new horizons.[6] Although Basque sheepherders were also often romanticized, they were a part of an industrial enterprise that responded to both national and international markets. Indeed, Basque immigrant workers can be seen as "Industrial Sheepherders" as Igler suggests for cowboys. The application of the word *industrial* divorces them from much of the romanticized interpretations created by popular contemporary sources that emphasize Basque laborers' devotion to the flocks and their heroic survival as lonely wilderness workers. Honing in on the industrial work of these Basque immigrant workers during this time, the rest of this chapter investigates the arrival of early Basque ranch interests in Nevada and their role in attracting future Basque sheepherders from the Old Country to do wage labor on their ranches.

Basque Pioneers from California Lay the Foundation for Nevada's Basque Sheepherding Tradition

In the 1850s, the gold rush in California brought some Basque families to the American West. They arrived primarily from the provinces of the northern Basque Country and Latin America, particularly from Argentina.

Typically, those early Basques had fled to South America during Spain's First Carlist War (1833–1839). After the late 1848 public announcement of the California gold discovery, Basques living in South America rushed to California to try their luck. Some later poured their work and capital from mining efforts into the livestock business, while others started their own business ventures in livestock ranching from the beginning, applying the lessons learned during their years in South America.

In the 1850s, California and its major metropolis, San Francisco, became the economic center of the Far West for many businessmen, traders, stockmen, and overall capitalist investors. Although it is difficult to determine how many Basques came to California in those years, by the early 1860s they became an influential group in San Francisco and the Central Valley. They ran primarily livestock ranch operations and other business ventures related to stock agriculture. All the Basque pioneering families emphasized mutual help to promote the welfare of their community. In the context of the rapid commercial and industrial expansion in California, these families amassed large holdings and became landowners by purchasing government land, and using their workers, friends, and family members as strawmen to purchase other extensive land parcels. The main goal was always to maximize profits and holdings in both land and stock.[7]

By the 1860s, the increasing populations and land prices in California caused stock operations to move more and more into marginal lands. The completion of the Transcontinental Railroad made it feasible to use distant marginal rangelands and to send the stock and wool to markets both to the West and to the East. The lands of faraway Nevada were examples of marginal lands now open for development and exploitation. The available public domain attracted many stock operators to Nevada. After 1859, the development of the Comstock Lode stimulated the expansion of livestock ranching in the Great Basin. During the 1860s and 1870s, some livestock operators, including Texas interests, saw the grazing potential of northeastern Nevada. Among these early livestock operations were the ranches built by Basques. By the 1860s, rural regions of Nevada increasingly became of interest to the western and eastern economic centers.[8]

Historian William Cronon contends that city and country were closely connected. The city acted as a nucleus and prod for the countryside's often distant development in terms of agricultural crops or stock enterprises. One lens that Cronon uses to analyze the correlation of the city with the surrounding rural areas is the economic model of geographical localization proposed by nineteenth-century German economist Johann Heinrich von Thünen who explained that the rent of land varies in concentric zones surrounding the urban centers. The closer the concentric zone to the city, the higher the land prices. The more distant the concentric belts, the less expensive the land, as befitted the usual marginal nature of the lands. These marginal lands of Nevada, which were well suited for extensive stock operations, were now connected to urban centers by the new railway transportation systems.[9]

Basques who settled in California in the 1850s and in the 1870s often extended their ranching businesses into the Great Basin. The arrivals of the Altube brothers, Jean Garat, the Arranbide brothers, Bernardo Ohaco, Antonio Harispuru, or Bernardo Yparre all marked the beginning of Basque immigration into this region. They provided the foundation for community attachment and solidarity. While they were permanent residents of California, they were often seasonal inhabitants of Nevada. The Silver State, however, increasingly became their operating center. Most owned their own homes in San Francisco and traveled occasionally to Nevada in order to supervise their enterprises. From the 1850s to the 1890s, the history of some early Basque ranch families, the Altubes and the Garats among others, can best be understood by following the different economic facets of their ranch operations.[10]

The Founding Basque Ranchers: The Altube Brothers and Jean B. Garat

In 1845, at an early age, Pedro Altube (1827–1905), who would later be known as the "Father of the Basques of the American West," immigrated to Argentina and joined his family there in the cattle ranching business. By 1850, following the gold boom, Pedro had settled in California. In 1851, his brother, Bernardo, joined Pedro in his economic ventures in California. They first started in the cattle marketing business. As the population of the Golden State increased, the two brothers found that market-oriented livestock production could be a more profitable business than mining. By 1855, as cattle prices dropped, the Altubes settled down in San Mateo to start a dairy farm. Other Basque immigrants and fellow Basques from Buenos Aires in Argentina, settled in the area of San Mateo as well. In the second half of the decade, after they had accumulated a small sum of money, the Altube brothers returned to the cattle business. They bought cattle in southern California at low prices and herded them north to the mining camps in the foothills of the Sierra. They and other Basques opened abattoirs, or slaughter houses, and butcher shops to expand their businesses. The Altubes also supplied cattle from their operations in the Central Valley to San Francisco.

While California was distant from the events of the Civil War during the 1860s, the decade saw upheavals in land prices and the stock industry, as well as weather crises. New economic conditions affected the Altubes' and other Basque ranches. The drought of 1863 and the flooding of the next year brought extensive losses of cattle. Some Basques were able to take advantage of high livestock prices arising from these weather events and apply their profits to buying newly available land in the Central Valley. In 1860, the Altubes owned nearly nine thousand acres of land and over one thousand head of livestock. During the following decade, the Altubes raised cattle and drove them on hoof to the San Francisco market. The Altubes succeeded in amassing land and capital to eventually finance a move of their livestock operations into northeastern Nevada.[11]

The uncertainty of livestock operations in California and competition with the Miller & Lux operations in the San Joaquin Valley in California pushed the Altube brothers and other Basques to find new rangelands across the Sierra in the high-desert environment of the intermountain Great Basin. In 1871, the Altubes settled in Independence Valley in western Elko County, Nevada. Throughout the first half of the decade, they progressively relocated their livestock business from California. It is widely acknowledged that the Altubes' operation was the first Basque outfit in Elko County. Situated forty-five miles northwest of Elko, the extensive new ranch lands had potential resources for mining and grazing enterprises. In the early 1870s, Elko County became the site of settlement by other ranchers from California and Texas herds. Within a short period of time, thousands of range cattle and sheep grazed the free and open ranges of the public domain. Private ranch properties were much smaller but were regarded as the base ranches from which these larger herds roamed about on the public domain.[12]

The Altubes formed the partnership called Spanish Ranch, which became one of the largest livestock companies in the Great Basin. In many ways they emulated the Miller & Lux operations in California. Since a large livestock operation was necessary for big profits, in order to consolidate their enterprise, they had to absorb many smaller scattered ranches and, of course, continue to exploit grazing resources on the public domain. During the 1870s, the Altubes acquired more than ten thousand acres, giving them key water rights and access to grazing on public lands. During the 1880s, the Altubes expanded their livestock business, exerting as much control as possible over public-domain grazing lands by monopolizing the sources of water through their water-right claims.

Despite their economic success, at the end of the decade, the situation for the Altubes' grazing operations faced a crisis. Like many other livestock operations in the northern Great Basin, the Altube brothers could not escape the disaster brought by the severe "White Winter" of 1889–1890. Although the winter almost ruined the Spanish Ranch, the ranch was slowly able to recover. Most free and open range–based operations faced upward of 90 percent loss of stock in the now desolated ranges of the Intermountain West. With hard work and perseverance, they were able to adopt a renewed competitive strategy in straitened times. Amidst liquidation proceedings by many stock operators, the Altubes continued land purchases and at the same time proceeded to combine sheep, cattle, and horse stock to respond to different market situations in the economically depressed decade of the 1890s.

By 1898, the Altubes' ranch, now called Palo Alto Land and Livestock Company, had recouped from the crisis. By the turn of the century, when the two brothers were in their fifties, they had amassed over seventy thousand acres, controlling all the resources in Independence Valley and consolidating their economic position in the livestock markets.[13] The Palo Alto Land and Livestock Company was recognized as the richest ranch in the northern Great Basin. The Altube family always maintained a residence in San Francisco and kept their properties in California. At the time of Pedro's death, the Basque ranch comprised seventy-five square miles in northern Elko County. In October 1907, the ranch sold to a syndicate composed of W.H. Moffat and H.G. Humphrey from Reno, and L.L. Bradley and Peter Garat from Elko.[14]

Like the Altubes, the Garat family was one of the early ranch operations that made a fortune in late nineteenth-century Nevada. The Garats were also Basques, who in the mid-nineteenth century followed the Atlantic triangular immigration pattern: Basque Country–South America–North American West. The same pattern of entrepreneurship that gave rise to the Altubes' livestock business also characterized the career of Jean Baptiste Garat in Nevada ranching.

In the first half of the 1850s, Jean Garat began raising cattle and selling beef to the mining camps in the foothills of the Sierra. In 1852, he and his wife, Gracianna, created the YP Ranch. The Garats started using the brand, which is considered one of the first livestock marks registered in the U.S. Patent Office. The Garats also engaged in the butchering business in Calaveras County. By the late 1850s, they began running cattle in the northern rangelands of the San Joaquin Valley.[15]

In the early 1870s, as with the Altubes, Jean Garat could not compete with Miller & Lux and moved out of California. Henry Miller pressured Garat to sell out his business. Miller imposed an exorbitant cost to water livestock and, although the Basque stockman resisted, finally, Garat had no other choice but to leave and find other ranges. He had heard about the potential grazing in northeastern Nevada. In 1871, he trailed about a thousand cattle eastward crossing the Sierra near Bridgeport, California, on to the Great Basin. The stock wintered near Lovelock Valley along the Humboldt River. They made temporary camps wherever they found water and good pastures. In 1874, the Garat family bought 320 acres of land from the Captain Stiles estate in White Rock in Elko County. Still, he did not break his connection with San Joaquin Valley and kept his California properties. The Garats also maintained their home base in San Francisco.[16]

The Garat & Company ranch bordered the property of the Spanish Ranch indicating an affinity among Basque immigrants in livestock ranching. Another Basque, Juan Baptiste Arrambide became a general partner with Garat in acquiring as much land as possible in Elko County. During the 1880s, the Garat ranch expanded rapidly. In 1882, Garat bought land from A.B. Chapman and a year later from J.B. Ringold. Jean Garat's stock wintered in the lower sub-basin of the Owyhee Desert rangelands. By the end of the decade, the Garats livestock company owned approximately ten thousand head of cattle.

The winter of 1889–1890 was disastrous for Garat & Company. The Garats lost over nine thousand head of cattle. Despite the advantages of sheep grazing in the Great Basin after the "White Winter," the Garats opted to continue with cattle as their major stock operation. During the 1890s, the Garats cooperated with other Basques in their struggle to recover from the economic losses. They also maintained cross-cultural relationships with non-Basque stock operators in Nevada. Indeed, in the early 1880s, Jean Garat was already a member of the Nevada Livestock Association. In 1895, the marriage alliance of Jean's son, John, and Matilda Indart served to consolidate business relationships between two Basque livestock families. Well into the twentieth century, Garat & Company continued to be run as a family business for four generations. It became one of the largest livestock companies in Nevada. In 1939, after John Garat died, the family sold the corporation for approximately $850,000, including seventy-five thousand acres of land, which then became the Petan Land and Livestock Company.[17]

The Beginnings of Basque Immigration

Although the Basque pioneering families did not constitute the first permanent settlement in Nevada, their economic ventures determined subsequent Basque migration and settlement in Nevada. Not surprisingly, the early Basque ranch operations served as the immigration conduit between the Basque Country and the American West, providing the first centers of employment for Basque workers. After 1869, the Transcontinental Railroad offered connections to the East Coast and facilitated Basque immigration to the Far West.[18]

The most significant period of the Basque immigration in terms of social and labor history occurred between 1890 and 1920. These early newcomers emigrated as part of the great wave of southern and eastern European immigrants in the United States often called the "New Immigration." The American industrial expansion benefited from access to international markets and a ready supply of immigrant labor. From the 1880s to the 1920s, the United States registered high immigration levels. These immigrants most often occupied the lower end of the occupational ladder in every sector across the country. The "New Immigration" profoundly altered the ethnic landscape in the United States. Basque immigration must be understood in the context of American economic expansion and the global economy that moved large numbers of workers to the United States.[19]

Sheepherding became the primary occupation open to Basque male immigrants in the American West. *Sheepherder* refers to the lower-ranking occupation in the sheep industry of the West. More precisely, sheepherders (or workers, laborers, or ranch hands) are the contract laborers employed by either oral or written agreement whose main duty is to care for and oversee the fattening of range sheep to be marketed for wool and meat. In the yearlong process, because of the arid and semiarid ecoregion, handling sheep on the western open range required grazing over great distances with sheep flocks of nearly two thousand head, constantly on the move from one pasture to another depending on the season. Novelist Mary Austin's description from her 1906 novel *The Flock* is still useful: "The sheep-herder is merely a hireling who works the flock in its year-long passage from shearing to shearing." The *hirelings* must be distinguished from the sheepmen who are the entrepreneurs and businessmen. In the hands of the sheepherder rested the marketable products of the business enterprise, both the investment and the prospect for future profits. The owners of the capital investment represented in the sheep risk a great deal as they entrust these animals to their caretakers. It was essential for owners—the sheepmen—and workers to develop trust and confidence in one another. But clearly the workers, "new immigrants" in a foreign land, stood at economic and power disadvantages in terms of negotiating their recompense and condition of work.[20]

Young male workers, typically single, provided the ideal workforce for sheepherding over extensive ranges. "New immigrants," according to historian Oscar Handlin, were "uprooted" from family, surroundings, and in these cases "transplanted" to the western ranges. The workforce was most often recruited from rural areas in the Basque Country with few skills beyond farming experience acquired in their small family farmsteads. They entered the sheep industry as laborers, having little knowledge of or familiarity with sheep. In the Basque Country some families owned only a few cattle and sheep, a situation that sharply contrasted with the American West ranges where bands of sometimes two thousand sheep grazed on vast public rangelands.[21]

The route from the Basque Country was a tiring and long journey. Traveling mostly to the ports of Bilbao, Bordeaux, and Le Havre, the prospective workers then faced a week of sea travel across the Atlantic and finally a four- to five-day train ride to the Far West. Arriving in the designated destinations, the young immigrants would jump off the train, some disoriented, and present themselves to their new employers. From this moment on, Basque immigrants were placed under the care and control of a ranching outfit. Sometimes, they found local Basque boardinghouses situated across from railway depots and waited there a couple of days until their employers or foremen arrived to conduct them to a distant ranch. The ranchers, in their position as employers, provided clothing and special equipment necessary to work in the rangelands. In most of the cases, the workers had to pay off these advances.[22]

After a couple of days, the ranchers placed these immigrants into a strange environment with a band of sheep counting about one or two thousand head. The new workers had to understand their new work duties and did not always understand the terms of their payment. Many felt at this point what one Basque official in Spain described as "the fraudulent promises which drives them to abandon their homes, emigrating to the American Republics" [writer's translation]. The new environment was unknown, as well as their job of driving sheep and adjusting to various aspects of American society. The loneliness in unfamiliar environments in the open rangelands of the Great Basin no doubt invoked alienation and sometimes despair in these immigrants. Their emotional reactions are difficult to reconstruct as they struggled to adjust themselves to a new world so different from the traditional life and ways of their home country. Sometimes, they expressed themselves in tree carvings or arborglyphs. They carved different types of figures, such as women shapes or patriotic symbols. Many of the aspen carvings made by Basque sheepherders have attracted the attention of scholars who see in them evidence of interaction with the natural world.[23]

Isolated work involving sheepherding in mountainous terrains under arduous weather conditions very often produced dangers. In late April 1915, a Basque sheepherder, Martin Goyeneche, got lost in the middle of a strong storm in Battle Creek, near Battle Mountain in Lander County, Nevada. He was in charge of a band of sheep on Beaver Creek. On April 28, on visiting the camp to locate him, the camp tender found the sheep without human guidance. The Basque immigrant worker was found dead a week after the incident.[24]

Because of the alienation and lonely work in desolate ranges and high mountains, there were some sheepherders who suffered from severe dementia, sometimes even being hospitalized in psychiatric facilities. When they went to town after long periods of work in remote rangelands and mountainous terrains, some acted like wild men, sometimes disturbing the peace and committing criminal acts, similar to cowboys after long cattle drives to railhead towns in the Great Plains.[25] For example, on May 2, 1898, Jose Miguel Rementeria, a Basque sheepherder, was brought by his employer John Guthrie to Winnemucca to receive medical assistance due to an accident. At the Lafayette Hotel, a doctor was called but could not arrive until the morning. During the night, the sheepherder apparently went mad and stabbed the Basque Hotel manager sixteen times. Although the critically-wounded man survived, Rementeria was arrested and later hospitalized at the state asylum in Reno.[26]

Because of the similar journeys many took to get to there, some might think that all the stories about the Basque immigrants entering the western sheep industry were the same. They were not, however, and despite the difficult and hard working conditions, many Basques succeeded in making a living and even prospering in the sheep industry, fulfilling many of their ambitions. The reasons behind emigration vary depending on each individual.[27] On the one hand, economic reasons prevailed. Like other European immigrants, the Basques emigrated from lands unable to support them. The nineteenth century witnessed crucial structural changes within the Basque Country, particularly in rural areas. The rapid industrial growth in the nineteenth century brought significant changes in traditional farming practices. The irruption and development of the capitalist mode of production altered the dynamics of agrarian production and way of life.[28]

Furthermore, the political convulsions of the nineteenth century, including two civil wars (the Carlist Wars), disrupted the old legal status of the Basque provinces.[29] After the Second Carlist War (1872–1876), the *fueros* system was abolished in the Basque Country. The *fueros* were a set of norms and customary rules that regulated community life in each of the Basque provinces, including the relations of these territories with the Spanish monarchy, and at the same time guaranteed the status of the local elites.[30] The socioeconomic and political changes that occurred during the nineteenth century became the main driving forces behind Basque emigration, particularly among the peasant population who had suffered most from these changes and the consequences of war. Thereof, emigration became an important expectation of many Basque young boys largely from rural areas.[31]

On the other hand, the Basque family structure itself pushed emigration. Besides the economic factors, family structure also created reasons for emigration. In the Basque Country, primogeniture meant that the eldest son was the inheritor of the family estate and in that position he took charge of all obligations. The eldest son entered into the possession of all the family estate, married, and had a large family. In the large Basque nuclear families, the prime heir stayed in the Old Country. While the system allowed family lands and the name to be perpetuated, it also denied other children (both male and female) titles or land ownership. For the latter, the choices could be: work for another house, join the priesthood or the army, marry other farmstead's heirs in the case of the women, or, mostly in the case of males, emigrate to America.[32]

Similar to the arrival of other immigrant communities in America, all of these factors established the classic "push-pull" model for migration when opportunities for skilled and unskilled labor occurred in the expanding economy of the United States in the late nineteenth century. The western American sheep industry offered something of a safety valve for those affected by the upheaval. While it did not promise great wealth, life as a sheepherder in Nevada attracted many Basque immigrants with few other opportunities at home, and news of some successful individuals encouraged more young men to make the journey.[33] Whether they emigrated owing to economic, political, or social reasons, all of them, in one way or another, went to America to find a new life.

The Recruitment and Organization of Work: Labor in the Sheep Industry

For many years, Nevada cattle barons, like their early counterparts in California, dominated the state's livestock agricultural affairs. In order to do this, they acquired land tracts strategically situated to control access to water sources to monopolize the public range; cooperated with local authorities to obtain political power and government assistance; and enjoyed willing and cheap labor from Mexican, Chinese, or European immigrants. Rising alongside this cattle industry, as Nevada sheep ranching became increasingly integrated into the national and international marketplace, it too sought the cheapest labor to keep the open-range sheep industry as profitable as possible and handle the rather unglamorous labor to herd and care for large bands of sheep. Much of the workforce for this labor came from the poorer regions of southern Europe, in other words, the Basque Country.[34]

As Nevada sheep ranching became increasingly integrated into the national and international marketplace, it sought the cheapest labor to keep the open-range sheep industry as profitable as possible. Despite the work's lack of glamor, the stock operations demanded a dependable, reliable labor force. And although this labor force, as Carey McWilliams once characterized agricultural labor in California, was an "agricultural proletariat," great responsibility was placed upon it to protect and preserve the lives of sheep that represented the overwhelming investment capital of the owners.[35]

Early Basque sheep ranchers quickly learned how these labor forces worked in the New World. To procure immigrant workers, Basque sheep operations found in the Old Country a large labor pool ready to fill these ranch jobs, often from their family and friends' circles. As the daughters of Ramon Lugea (1874–1958) remembered, their father came to Elko in 1908 where he worked first as a sheepherder for the Spanish Ranch. Others from the Basque Country often took their first jobs with the Altubes or the Garats. They formed an immigrant working class that made up the sheepherding labor pool. Usually, immigrants were recruited in informal ways. For instance, private personal letters from immigrants urged and encouraged their family members and friends to emigrate and work as sheepherders.[36]

By the 1890s, however, the early Basque immigrants in the sheep industry started to make the pattern of the "agricultural ladder" work for them and subsequent newcomers.[37] The following stages were distinguishable: unpaid or poorly paid work; compensation being made in lambs; operating mixed sheep bands (of both owners and sheepherders' lambs); becoming autonomous graziers (oftentimes forming partnerships); investing in an operation by purchasing land and property. Of the early stages on this ladder, receiving wages in lambs could be important because it enabled many Basque immigrants to prosper and expand their own operations. Among those who started as a laborer and moved up the ladder was Charles Garteiz, a Basque youth who, in 1892, emigrated at the age of fourteen to the United States and settled in Humboldt County, Nevada, traveling overland on the Transcontinental Railroad from the East Coast. He began working as a ranch hand, driving cattle for the company owned by Pick Anderson. Some years later, still at a young age, he entered the sheep business when he formed a partnership with his brothers, Frank and Pete. They invested in the sheep industry, applying the lessons learned from earlier Basque sheepmen. The family partnership continued until 1921, when finally they sold the business. Later, in 1926, the Garteiz Brothers bought the Miller & Lux properties at Bilk Creek, near Denio, Nevada, and another ranch near Winnemucca. They invested in infrastructure, buying agricultural machinery, claiming and buying land, and also securing a labor supply among their relatives and friends from the Basque Country.[38]

Basque ranchers were in a position to use cultural ties to profit from other Basque immigrant labor more than other non-Basque ranchers. This can especially be seen in the recruitment process where family and personal contacts persuaded workers to undertake the long journey to America, only to face precarious working conditions. Sheepherders needed guidance from the very beginning, even before they departed to the New World. Before their departure, sometimes a sheep company had even already contracted with them and given them instructions to reach their assigned destinations. Employers advanced immigrants' travel costs, which the worker had to pay back fairly quickly. Sometimes the immigrants' families used their savings to pay for their sons' tickets. Although the cost of buying both a steamship and train tickets were expensive, sending a child to America to work as a sheepherder for a relatively short period of time was considered an investment. Sometimes, poor Basque rural families made severe financial sacrifices to purchase their sons and daughters' tickets to the American West. In this way, an emigrant was able to save money and bring labor earnings home after an intense four- or five-year working stint. If for one reason or another the immigrant could not continue living under those working conditions in the New World, he could always return to his country.

The language barrier left Basque immigrants vulnerable to various misunderstandings, particularly when dealing with non-Basque employers. For example, they knew little of anything about negotiating effectively with their non-Basque employers. Their employers essentially determined the conditions of their employment as well as their wages and provisions needed to survive in the new environment and even provided legal advice. The result was a mutual dependency of a master-servant-like relationship that often characterized the sheepherding working environment. Since sheepherders spent most of the year tending sheep in remote places, they became dependent upon their foremen as trusted patriarchal advisers and supervisors. Among other things, the employers secured the workers' savings in the bank, and mediated and corresponded with their families and friends in the Old Country directly.[39]

As a result of this employment structure, when Basque workers came, they were often at the mercy of unfair employers who took advantage of them, sometimes including Basque ranchers themselves. In the absence of fair labor standards in either state or federal laws, livestock companies in Nevada did what they pleased in labor relations. Large stock companies were not usually interested in formalizing rules about labor relations. Stock ranchers preferred to protect their autonomy and resisted any regulation of their relations with labor. The unionization going forward in Nevada mining communities with the state's new twentieth-century mining boom was irrelevant to ranch labor and operations.[40] In 1915, the Nevada Legislature established the Nevada Labor Commission to oversee labor relations. The Commission, however, had little interest in labor on livestock ranches.[41]

"The Brave Basque": Racialized Views of the Basques in the United States

At the same time that Basque immigrants entered the western sheep industry in the second half of the nineteenth century, several European anthropologists went to the Basque Country with the intention to learn more about this land and to collect information about its inhabitants. Their publications portrayed Basques as an ancient, anomalous, and quixotic race distinct from any Indo-European people. At that time, race meant both physical attributes (for example, skull shape and size) and sociocultural traits. While there was a wide range of speculation about Basques and their origins, the new scholarship produced a rather idealized and even museumized image of the Basques. These ethno-racial hypotheses did not answer questions on the origins of the Basques. The origin mystery added to the image of the romanticized Basque.[42] Such depictions of Basques as represented in these anthropological studies ultimately attracted the curiosity and the interest of the dominant society in the American West, paving the way for the acceptance and integration of the Basques.

By the 1870s, the word "Basque" had started appearing frequently in the American newspapers. Articles even echoed the contemporary anthropological interpretations about the Basques.[43] In 1875, the *New York Times* published a report on the ethnographical observations of a special correspondent who went to Hendaye (Labourd) in the northern Basque Country to learn more about the Basque culture. The author found a curious mixture of festivals and Catholic religious life in the Basque Country. In addition, the American traveler noted a distinctive Basque racial identity and attractiveness in physical appearance: "The type of the Basque race, both male and female, is not molded on a large scale, but the women, although *petite*, are well proportioned, and if not positively handsome, they are, at least, healthy-looking, and have not that sallow complexion which is characteristic of the male sex through Spain."[44]

In the summer of 1878, a correspondent of the *Cleveland Leader* traveled to the Basque Country to learn more about the local people. To him, everything was enchantment and exotic. He introduced the Basques in the following manner: "At this stage of my journey there seemed no fear that I should lack for food or shelter. I was fairly [*sic*] in the Basque country, and knew that this people were among the most hospitable of races." Travel through the Basque Country piqued the curiosity of this reporter about the local people, and this led him to further questions about this "strange race." His view of the Basque racial type was based on both physical qualities and behavioral traits. "As my attention fastened itself upon their frank healthy faces, and lithe, muscular frames—each, according to his years, being a type of manly beauty and symmetry—," this correspondent wrote, "I was struck with admiration for the race from which they spring...one whose origin is lost in the night of antiquity, still holding to its native fastnesses, intact in its usages and language, and absolutely uncorrupted in its tribal identity. This reporter noted a racial identity that put together the Basques' alleged natural qualities and temperamental attitudes, especially: valor, hospitality, goodness, kindness, determination, trustworthiness, integrity, sovereignty, fastness, and purity.[45]

By the 1880s, public opinion commonly cast Basques as an exceptional race. In 1882, the *New York Times* published an article on the Basque fishermen and their longtime connection to the sea since the days of fishing and whale hunting. The article used "brave, honest, and industrious," as the main adjectives in describing the Basque people, "while both men and women are always cheerful and light-hearted." The text described the importance of the fishing industry for the Basques, by saying that "they belong to a people who, for centuries, have repelled foreign invasion, have enjoyed free institutions, and made their own laws. The Basque fishermen are the descendants of the old whalers, and retain their traditions."[46]

In the 1890s, testimonies about the Basque people continued to appear in American newspapers. In 1893, the *New York Times* reprinted a small article from the *London Globe* on the uncertainty about the origins of the Basque language. It was titled: "Basque Is a Lonely Tongue Still." In June of 1896, the *New York Times* reprinted an article from *The Gentleman's Magazine*, which asked a persistent question, "Who Were the Basques?" Echoing the theories of contemporary scholars, the physical anthropologist Paul Broca noted that the Basque crania were distinguished from the African and European type. Just two months later, in the same newspaper an article took up the subjects of the Basques during the advance of the Roman Empire into Western Europe and the strategic position that the Basque Country occupied and its importance to Rome. The Basques had the reputation of standing as "a rock against Roman and all succeeding waves [of conquest]." This idea was common among many American interpreters who considered Basques the "Warders of the Pyrenees."[47] But, as historian Coro Rubio has explained, this was largely an image of the Basques that had been constructed by the political elites of the Basque Country decades earlier.[48] These depictions of the Basques, published in papers throughout the United States, would contribute to how perceptions of Basque immigrant sheepherders took shape in the Far West.

The development of the Spanish-American War would also lead to further articles of this kind. In 1898, while the United States and Spain were negotiating peace, an American newspaper published a favorable notice about the Basques.[49] The title itself speaks positively of the Basques: "Valiant Vizcayans: Spaniards Who Are Always Against the Government." They were, according to the text, "overrun, but never conquered." The article described generally how the Basques from the province of Biscay fought in the Second Carlist War (1872–1876) to defend and protect their "*fueros*." The newspaper article stressed that "these *fueros* have been a source of infinite annoyance to the government from the fact that the Vizcayans form thus an independent community in the kingdom and have privileges so exaggerated that the people of other provinces complain bitterly of the immunities enjoyed by the Basques." As this article explained, the conclusion of the Second Carlist War in 1876 brought, among other things, the abolition of the *fueros*. The author contended that Basques were proud of their distinctive privileges. The author also contended that the Basques were good warriors because they learned how to adapt to their environment: "These mountains are the natural habitat of the Basques, and whenever an insurrection breaks out the inhabitants remove to the mountain fastnesses and can indefinitely prolong the war against any force brought to bear upon them." The Basques, it concluded, "are mountaineers, and mountaineers in every country are independent, liberty loving and impatient of any form of political control save when exercised by their own people." Furthermore, "they have never patiently submitted to a foreign yoke."[50]

The article included a cartoon showing a supposed "Vizcayan warrior" in a Basque village, holding a shotgun and wearing what it supposed to be traditional clothing. However, rather than a Basque native soldier, this drawing actually suggested an image of a Moorish warrior of Northern Africa, with the classic turban and baggy wear. The Orientalized image unwittingly displayed the Basques as members of a Mediterranean race. Under the whiteness paradigm in the United States, this image could lead an American reader of the time to see Basques as coming from a non-White "race" of southern Europe. His darkened face made him strange and angry, but at the same time it portrayed a fierce protector of his homeland. Moreover, although the article is talking about the Vizcayans (Basques from the province of Biscay), the reader might understand that the author was actually thinking in terms of all the Basque territories.[51]

Along with the Catholic faith, the *fuero* was one of the most important components of the nineteenth-century Basque identity.[52] In the course of the nineteenth century, furthermore, according to historian Coro Rubio, Basques forged a singular identity to differentiate themselves from the rest of Spain. Rubio suggests that a Basque elite invoked this exceptionalism to secure their political and economic status (*fueros*).[53] In the same vein as Coro Rubio's work, I further suggest that American public opinion reinforced this belief, which made Basques socioculturally, politically, and racially exceptional in the United States. As such ideas gained currency across the United States, they contributed to a high regard for the Basque immigrant sheepherders in the American West. Other newspapers in the American West would repeat these depictions at the same time that the Basque sheepherder population was appearing in that region.

Generally, newspapers depicted Basques superficially, with a standardized version about Basques and their identity. On August 3, 1899, an article in a Nevada newspaper appeared with the following descriptive title: "The Brave Basque." According to this text, Basques were "proud and unconquered people." It further wrote: "Only in this tiny [Basque] country, among the mountains have these sturdy, stalwart people succeeded in retaining anything of their own individuality."[54] And this apparent individuality, culturally constructed, made them stand out as an exceptional racial stock in the American public imagery.

As writers continued to depict the Basque as noble survivors of long centuries of struggle, the strength of the Basque character within the sheepherding industry began to also emerge. For example, on December 14, 1901, an American journalist wrote, "the best sheepherder in the world is the Basque." The text speculated that only Basques were psychologically capable to confront the isolation demanded of sheepherders in the most remote rangelands of the American West:

> No matter how strong the man may be at the start, if he will follow sheep long enough he will wind up a driveling idiot. No man has ever been able to explain why this is, but it is a fact, nevertheless. I have known young, healthy Mexicans and Americans to begin herding sheep, and in less than five years their brains had given way and they could not talk coherently. Perhaps there is something in the theory which is sometimes advanced that constant association with the sheep reduces their intellectual capacity to the level of their associates, or perhaps there is something in the theory that the loneliness and the solitude of the calling will wear away the brain tissue of any man.[55]

In working with the sheep, according to this article, Basques were better able to stand the loneliness and "the strain than other nationalities." Next to the Basques in endurance were the Mexicans. Both, stated the text, "can stand following the sheep for years, but both will succumb in time. An American [can] hardly stand the strain for a year, and I have know[n] some of them who did not stand it that long." In closing, the article claimed: "There are thousands of insane Mexicans on the border today who became so by reason of their occupation as sheepherders and for no other."[56] The essentialist view that attributed to Basques special sheepherding skills, as we will see later, also conferred upon them a degree of racial whiteness and values that entitled them to a privileged labor category.

Equating Basques and people of a good European racial type was commonplace among contemporary observers in the popular press. These perceptions prevailed in American public opinion for years. In

1916, for example, the Missourian *St. Louis Globe-Democrat* published an article under the title: "Basques of the Pyrenees." The story contended that Basques were "the pure aborigines of Europe." In 1918, another American newspaper wrote "Basque Race Are a People of Mystery," while stating that "they are a proud race, proud of their ancient traditions."[57]

Nevertheless, by the turn of the century, although Basques assumed the status of a curious race that was brave and independent—an image that would continue to be used by sheep operators for decades in the advocacy for additional Basque immigration—Basque were also oftentimes discriminated against. The debate about instituting those regulations frequently attempted to use racialized language to depict the Basque sheepherders as an "Other" for political and profit reason. As the conservation movement gained momentum by the end of the nineteenth and early twentieth centuries, the U.S. Forest Service, chief representative of that movement in the West, imposed range regulations that oftentimes excluded sheep and favored instead the grazing of cattle in high mountain pastures.[58]

Summary

Basque immigrants began to arrive in significant numbers in the United States in the second half of the nineteenth century. The late 1870s saw the beginning of a wave of Basque immigrants to the Great Basin seeking opportunities in the open-range sheep industry. Many of these immigrants were the younger sons of peasant families in the Basque Country—those who were not in the line of inheritance in their family. These Basque immigrants, while encountering unfamiliar and difficult labor conditions, would become sheepherders in Nevada and the West, ultimately forming the base of the western sheep industry labor force.

As Basque immigrants set out upon their diaspora into the American West, academic studies of late nineteenth-century physical anthropologists racialized the Basque people, attributing to them potential distinguishing racial features. Such studies created a Basque racial identity that would be echoed by the popular press.

Racialized depictions of the Basques would continue over the coming decades to affect the way Basque immigrants were perceived, ultimately providing both opportunities for Basque immigrants while also leading to instances of marginalization. Despite the existing positive cultural stereotypes, older cattlemen would scapegoat Basque workers for all the problems associated with grazing on public-domain lands.

The next chapter analyzes how the Basque presence as transient sheepherders in Nevada and the vast herds of sheep they tended not only created an agenda for conflict that marked the Nevada range struggles, but also influenced Forest Service decisions about who should graze and have access to high-mountain pastures.

Notes

1. Interview, Pedro "Pete" Barinaga, August 20, 1975, transcript, Oral Histories, Basque Museum and Cultural Center, Boise, Idaho; obituary, "Pete Barinaga Sr." *Idaho Statesman*, January 14, 1985.
2. On the same steamship, from Markina travelled: Leocadio Arrate, Pedro Bilbao, Luis Aspiri, Jose Churruca, Jose Maria Ibarlucia, and Francisco Ibarlucia. And from the Basque province of Guipúzcoa were: Donato and Jose Manuel Araquistain from Deba; Aniceto Garate from Mutriku; Jose Manuel Badiola from Zestoa. Declaration of Intention, Pedro Barinaga, no. 19 2112, April 9, 1928, State District Court, Elko County Courthouse, Elko, Nevada; Certificate of Arrival in United States 1911, Pedro Barinaga, no. 19 2112, August 5, 1933, U.S. Department of Labor, Bureau of Naturalization, State District Court, Elko County Courthouse, Elko, Nevada; Petition for Citizenship, Pedro Barinaga, no. 356, September 17, 1934, State District Court, Elko County Courthouse, Elko, Nevada; Passenger Record, Pedro Barinaga, March 14, 1911, New York Passenger Arrival Records, 1820–1957, U.S. National Archives, New York; Barinaga Oral History, Basque Museum.
3. For a succinct explanation of Valentin Aguirre's transitory hotel, see Echeverria, *Home Away from Home*, 45–47.
4. Barinaga Oral History, Basque Museum; "Requiem Said for Nevadan," *Nevada State Journal*, April 19, 1952, 18.
5. White, *"It's Your Misfortune and None of My Own,"* 226; Sawyer, *Nevada Nomads*, 163–74.
6. Igler, *Industrial Cowboys*, 7.
7. Azcona Pastor, *Possible Paradises*, 230–240; Douglass and Bilbao, *Amerikanuak*, 203–33.
8. Pisani, *To Reclaim a Divided West*, 183–188; John M. Townley, *Turn This Water Into Gold: The Story of the Newlands Project* (Reno: Nevada Historical Society, 1998), 1–9.
9. Cronon, *Nature's Metropolis*, 48–52.
10. Marie-Pierre Arrizabalaga, "A Statistical Study of Basque Immigration into California, Nevada, Idaho, and Wyoming between 1900 and 1910" (master's thesis, University of Nevada, Reno, 1986), 36–41; Douglass and Bilbao, *Amerikanuak*, 204–38.
11. Carol W. Hovey, "Pedro and Bernardo Altube: Basque Brothers of California and Nevada," in *Portraits of Basques in the New World*, eds. Richard W. Etulain

and Jeronima Echeverria (Reno: University of Nevada Press, 1999), 57–63; Pedro Altube, Stockman, Vertical Files, Nevada Historical Society; Patterson, et al., *Nevada's Northeast Frontier*, 230–32, 387–91; Edna B. Patterson "Spanish Ranch Letters," *Northeastern Nevada Historical Society Quarterly* 84, no. 1 (Winter 1984): 3–15; Elliott, *History of Nevada*, 120–21.

12. On April 23, 1870, the *Elko Independent* appeared very optimistic about the county's economic future, especially in grazing and mining, because of both its plenty natural resources and the strategic positioning in terms of market structure: "The supply of the grass essential, an abundance of water, is inexhaustible in this locality. The fertile valley below the placer ground, the rich quartz veins on the mountains above, and its accessibility from the present continental railroad by an easy and well constructed wagon road, will soon bring to its rugged mountain an army of hardy miners, who will dig out its buried wealth and cultivate the fertile valleys." *Elko Independent*, April 23, 1870.
13. In late 1901, Barney Horn, a Californian cattleman, bought 2,500 head of cattle from the Altube Ranch, which were sent directly to the San Francisco market. Horn paid $40 apiece for steers and $30 for heifers. *Times-Review*, November 5, 1901.
14. Hovey, *Portraits of Basques in the New World*, 64–79; Pedro Altube, Vertical Files; Patterson et al., *Nevada's Northeast Frontier*, 389–91; Patterson, "Spanish Ranch Letters," 7–15; "Livestock Interest. A Review of the Situation in Elko County," *Reno Weekly Gazette and Stockman*, February 27, 1890, 3; Carey McWilliams, *Factories in the Field: The Story of Migratory Farm Labor in California* (Boston: Little, Brown and Co., 1939), 28–39; Patterson, "Early Cattle in Elko County," 9–10; "Elko Stockman Dead," *Reno Evening Gazette*, August 10, 1905, 1; "Spanish Ranch Bought for a Million of Dollars," *Nevada State Journal*, October 8, 1907, 1.
15. Patterson et al., *Nevada's Northeast Frontier*, 217–18; Patterson, "Early Cattle in Elko County," 9; Hovey, *Portraits of Basques in the New World*, 61–64; Douglass and Bilbao, *Amerikanuak*, 256–57; Shawn Hall, *Old Heart of Nevada: Ghost Towns and Mining Camps of Elko County* (Reno: University of Nevada Press, 1998), 81.
16. See also Mary Catherine Miller, *Flooding the Courtrooms: Law and Water in the Far West* (Lincoln: University of Nebraska Press, 1993).
17. Patterson et al., *Nevada's Northeast Frontier*, 230, 418–22; Hovey, *Portraits of Basques in the New World*, 66–75; Douglass and Bilbao, *Amerikanuak*, 257–61; Hall, *Old Heart of Nevada*, 81; "Pioneer Nevada Stockman Dies in S.F.," *Reno Evening Gazette*, August 18, 1939, 3.
18. Craig Campbell, "The Basque-American Ethnic Area: Geographical Perspectives on Migration, Population, and Settlement," *Journal of Basque Studies* 6 (1985): 83–89.
19. Hatfield, "'We Were Not Tramp Sheepmen,'" 28–48; Carl Wittke, *We Who Built America: The Saga of the Immigrant* (New York: Prentice-Hall, 1940), 405–9.
20. Moroni A. Smith, *Herding and Handling Sheep: On the Open Range in U.S.A.* (Salt Lake City, Utah: n.p., 1918), 3–16; Mary Austin, *The Flock* (Boston: Houghton Mifflin Co., 1906), 56; Randall Howard wrote: "The sheep-herder, in his strange, lonely life on the range, almost wholly beyond the reach of civilizing influences,

is not thought of as a wealth-protector or a wealth-producer. Yet this little known, less understood sheep-herder of the interior is a vital link in a great Western Industry. He is the sole protector, during the greater part of the year, of an estimated 31,000,000 sheep, valued at $108,500,000. It is doubtful if any group of workers in the West have such a large amount of wealth so exclusively under their control for such long periods of time as do sheep-herders." Randall R. Howard, "The Sheep-Herder," *Outlook* 96 (September/December 1910): 943.

21. Oscar Handlin also wrote that "emigration took these people [new immigrants] out of traditional, accustomed environments and replanted them in strange ground, among strangers, where strange manners prevailed." Oscar Handlin, *The Uprooted: The Epic Story of the Great Migrations that Made the American People* (Boston: Little, Brown and Co., 1951), 62, 257, 5; Edlefsen, "A Sociological Study of the Basques of Southwest Idaho," 51–53.
22. Patrick Bieter, "Reluctant Shepherds: The Basques in Idaho," *Idaho Yesterdays* 1 (Summer 1957): 10–15; Bieter and Bieter, *Enduring Legacy*, 9–16, 30–41.
23. José Colá y Goiti, *La Emigración Vasco-Navarra* (Vitoria: Diputación de Álava, 1882), viii; Edlefsen, "Sociological Study of the Basques of Southwest Idaho," 51–53; Richard Lane, "Basque Tree Carvings," *Northeastern Nevada Historical Society Quarterly* 1, no. 3 (Winter 1971): 1–9, see also Jose Mallea-Olaetxe, *Speaking through the Aspens: Basque Tree Carvings in California and Nevada* (Reno: University of Nevada Press, 2000).
24. Richard H. Lane, "Trouble in the Sweet Promised Land: Basques in Early Twentieth Century Northeastern Nevada," in *Anglo-American Contributions to Basque Studies: Essays in Honor of Jon Bilbao*, eds. William A. Douglass, Richard W. Etulain, and William H. Jacobsen, Jr. (Reno: Desert Research Institute Publications on the Social Sciences, 1977), 33–34; "Basque Shepherd Lost in Storm," *Reno Evening Gazette*, May 4, 1915, 3.
25. Robert R. Dykstra, *The Cattle Towns* (New York: Alfred A. Knopf, 1968), 112–48. For a new interpretation of the cowboys of the American West, see: Igler, *Industrial Cowboys*; Jacqueline M. Moore, *Cow Boys and Cattle Men: Class and Masculinities on the Texas Frontier, 1865–1900* (New York: New York University Press, 2010).
26. Lane, *Anglo-American Contributions to Basque Studies*, 34; *Idaho State Journal*, February 20, 1976; Louis Irigaray and Theodore Taylor, *A Shepherd Watches, A Shepherd Sings* (Garden City, New York: Doubleday and Co., 1977), 21; "Frank Germain Assaulted," *Nevada State Journal*, May 5, 1898; probably the most notorious case of a Basque sheepherder's insanity was Domingo Echeverria, who in September of 1951 murdered a nurse, Elizabeth Catlett, at the Humboldt County Hospital while receiving medical attention. Echeverria was eventually convicted of murder and executed, *The State of Nevada v. Domingo Echeverria*, case no. 319, 1951, Trial Records, County Clerk Office, Humboldt County Courthouse, Winnemucca, Nevada; "Echeverria Sentenced to Gas Chamber," *Humboldt Star*, November 13, 1951; on the social world of the Basque sheepherders in the American West, see also: Iker Saitua, "*Distilling Spirits*: Inmigrantes vascos, cultura de la bebida y prohibición en el estado de Nevada, 1910–1920," *Historia Social* 1, no. 90 (2018): 45–65.

27. A collection of articles that examine the Basque emigration to America, particularly to Latin America, is: Alberto Angulo Morales and Óscar Álvarez Gila, coords., *Las migraciones vascas en perspectiva histórica (siglos XVI–XX)* (Bilbao: Universidad del País Vasco, 2002).
28. Douglass and Bilbao, *Amerikanuak*, 131; Eduardo Jorge Glas, *Bilbao's Modern Business Elite* (Reno: University of Nevada Press, 1997), 92–104, 212–13. On the development of industrial capitalism in the Basque Country, see: Manuel González Portilla, *Estado, Capitalismo y desequilibrios regionales (1845–1900): Andalucía, País Vasco* (San Sebastián: Haranburu/Universidad del País Vasco, 1985); Manuel González Portilla, Manuel Montero García, José María Garmendia Urdangarín, Pedro A. Novo López, and Olga Macías, *Ferrocarriles y desarrollo: Red y mercados en el País Vasco (1856–1914)* (Bilbao: Servicio Editorial de la Universidad del País Vasco, 1996).
29. See: Coro Rubio Pobes, *Revolución y tradición: El País Vasco ante la Revolución liberal y la construcción del Estado español, 1808–1868* (Madrid: Siglo XXI, 1996).
30. Coro Rubio Pobes, *Fueros y constitución: La lucha por el control del poder, País Vasco, 1808–1868* (Bilbao: Universidad del País Vasco, 1997).
31. Douglass and Bilbao, *Amerikanuak*, 130–45; Glas, *Bilbao's Modern Business Elite*, 92–96; William A. Douglass, "Rural Exodus in Two Spanish Basque Villages: A Cultural Explanation," *American Anthropologist* 73, no. 5 (October 1971): 1109. To learn more about the social, political, and economical forces that propelled Basque emigration in the late nineteenth and early twentieth centuries, see José Luis de la Granja, Santiago de Pablo, and Coro Rubio Pobes, *Breve Historia de Euskadi: De los fueros a la autonomía* (Barcelona: Debate, 2011).
32. María Pilar Pildain Salazar, *Ir a América: La emigración vasca a América, Guipúzcoa, 1840–1870)* (San Sebastián: Sociedad Guipuzcoana de Ediciones y Publicaciones, 1984), 17–19.
33. Pildain Salazar, *Ir a América*, 11–19; Hatfield, "'We Were Not Tramp Sheepmen,'" 38–48. The German immigrant experience in the Great Plains had some similarities to that of the Basques. See: Frederick C. Luebke, *Immigrants and Politics: The Germans of Nebraska, 1880–1900* (Lincoln: University of Nebraska Press, 1969).
34. David Igler, "Industrial Cowboys: Corporate Ranching in Late Nineteenth-Century California," *Agricultural History* 69, no. 2 (Spring 1995): 201–15; Young and Sparks, *Cattle in the Cold Desert*, 233–35; Sawyer, *Nevada Nomads*, 162–73.
35. Elliott, *History of Nevada*, 121; McWilliams, *Factories in the Field*, 48; historian Gunther Peck discusses early Basque shepherd laborers working under a system of "tribute" or being paid in kind rather than in wages. The free and open ranges were marginal lands in Nevada that made possible the grazing of immense herds over extensive distances, but also required close supervision by a laboring class that worked for very little recompense either in wages or in kind. It can be ventured to say that marginal lands attracted and even required marginal labor. According to Peck, the constriction of public rangelands with the coming of national forests and national parks forced employers to move toward wage compensation to labor. Attributing the rise of wage labor in the sheep industry

to the diminishment of free land and forage needs further investigation. Peck, *Reinventing Free Labor*, 34–35.

36. Etulain, *Basques of the Pacific Northwest*, 1–4; Interview, Elena Lugea Arregui and Marianne Lugea Goicoechea, June 2, 1998, transcript, Oral History Program, Northeastern Nevada Museum, Elko, Nevada.
37. W. J. Spillman, "The Agricultural Ladder," *American Economic Review* 9, no. 1 [Suppl.] (March 1919): 170–79; see also John T. Schlebecker, *Whereby We Thrive: A History of American Farming, 1607–1972* (Ames: Iowa State University Press, 1975); David B. Danbom, *The Resisted Revolution: Urban America and the Industrialization of Agriculture, 1900–1930* (Ames: Iowa State University Press, 1979).
38. Lane, "The Cultural Ecology of Sheep Nomadism," 131–38; "Death Takes C. Garteiz at Home Here," *Humboldt Star*, October 19, 1951, 1; "Garteiz Ranch Interest Sold," *Reno Evening Gazette*, July 21, 1945, 3; in 1884, William Shepherd did not make distinction between sheepherder and owners when referring to the immigrant sheepherders in the California's Sierra: "The land is, however, owned, and if not fenced, is constantly ridden over by the boys, who drive off outside cattle and carry on a perpetual warfare with the Basque and Portuguese owners of bands of sheep which have to traverse the ranges on the way to the mountains or to the railroad." William Shepherd, *Prairie Experiences in Handling Cattle and Sheep* (London: Chapman and Hall, 1884), 121–23.
39. Pastor, *Possible Paradises*, 364–84; Douglass and Bilbao, *Amerikanuak*, 262–68; Bieter and Bieter, *Enduring Legacy*, 39–41; Barinaga Oral History, Basque Museum.
40. Russell R. Elliott, *Nevada's Twentieth-Century Mining Boom: Tonopah, Goldfield, Ely* (Reno: University of Nevada Press, 1966), 251–72; Sally Zanjani, *Goldfield: The Last Gold Rush on the Western Frontier* (Athens: Swallow Press/Ohio University Press, 1992), 46–48, 56–66.
41. *Second Biennial Report of the Commissioner of Labor of the State of Nevada, 1917–1918* (Carson City: State Printing Office, 1919), 7–10.
42. Rosario Calderon and Esther Rebato, "Historia de la antropología biológica en el País Vasco," *Revista Internacional de los Estudios Vascos* 42, no. 1 (1997): 51–52; Jesús Azcona Mauleon, "La delimitación antropológica y etnológica de lo vasco y de los vascos," *Cuadernos de etnología y etnografía de Navarra* 14, no. 40 (1982): 754–73; Joseba Zulaika, *Del Cromañón al carnaval: Los vascos como museo antropológico* (San Sebastián: Erein, 1996), 47–57; William Z. Ripley, *The Races of Europe: A Sociological Study* (New York: D. Appleton and Company, 1899), 180–204; Rodney Gallop, *A Book of the Basques* (London: Macmillan, 1930), xv, 282, 9, 1, 283.
43. See, for example: "Fossil Human Races," *New York Times*, April 28, 1874, 12; "Scenes in Spain," *New York Times*, January 10, 1875, 2.
44. "A Basque Holiday," *New York Times*, February 15, 1875, 10.
45. "The Basque People," *New York Times*, August 21, 1878, 3.
46. "The Early European Whalers," *New York Times*, March 19, 1882, 12.
47. "Basque is a Lonely Tongue Still," *New York Times*, October 8, 1893, 17; "Who Were the Basques?" *New York Times*, June 28, 1896, 26; "To Follow a Roman Road," *New York Times*, August 28, 1896, 13; "The Basque Provinces," *New York Times*, September 3, 1899, 12.

48. Coro Rubio Pobes, "Centinelas de la Patria: Regionalismo vasco y nacionalización española en el siglo XIX," *Historia contemporánea* 2, no. 53 (2016): 393–425.
49. It should be noted that later on, in May 1902, Sabino Arana Goiri on behalf of the Basque Nationalist Party (EAJ-PNV) sent a telegram to U.S. president Theodore Roosevelt congratulating him on the military victory over Spain in Cuba. Before this telegram reached America, the civil governor of Biscay intercepted it and considered it an offense against the Spanish government. Then, Arana was imprisoned and prosecuted, being finally absolved. Theodore Roosevelt himself asked the Spanish government to explain why it forbade Arana from sending a telegram to him. The case of Arana was widely covered in the American press. Alexander Ugalde, coord., *Patria y libertad: Los vascos y las guerras de independencia de Cuba 1868–1898* (Tafalla: Txalaparta, 2012), 259–61; "Spain Asked to Explain?," *New York Times*, May 31, 1902, 9; *Des Moines Daily Leader*, May 31, 1902; *Norwalk Evening Herald*, September 26, 1902; "Senor Arana Acquitted," *New York Times*, November 9, 1902, 5; *Syracuse Post Standard*, November 9, 1902; *Palo Alto Reporter*, November 13, 1902; *Chariton Democrat*, November 13, 1902; *Pocahontas County Sun*, November 13, 1902; *Algona Courier*, November 14, 1902.
50. "Valiant Vizcayans: Spaniards Who Are Always Against the Government," *Daily Ledger*, August 11, 1898.
51. Ibid.
52. Coro Rubio Pobes, *La identidad vasca en el siglo XIX: Discurso y agentes sociales* (Madrid: Biblioteca Nueva, 2003), 99–123.
53. Rubio Pobes, *La identidad vasca en el siglo XIX*. See also her article: "Centinelas de la Patria," 417–18.
54. "The Brave Basque: A Proud and Unconquered People Who Live in the Mountains of Spain," *The News*, August 3, 1899, 2.
55. "Sheepherders Go Insane," *Des Moines Daily News*, December 14, 1901, 4; the article was reprinted in several American newspapers, including: *Indianapolis Journal*, November 17, 1901, 2.
56. *Des Moines Daily News*, December 14, 1901, 4.
57. "Basques of the Pyrenees," reprinted in the *Indiana Evening Gazette*, April 28, 1916; "Basque Race Are a People of Mystery," *New Castle News*, July 31, 1918, 15.
58. Samuel P. Hays, *Conservation and the Gospel of Efficiency: The Progressive Conservation Movement, 1890–1920* (Cambridge: Harvard University Press, 1959), 54.

PART II

The Struggle for Legitimacy

3

Encroaching upon Forbidden Ground

Basque Immigrant Sheepherders and the Creation of National Forests in Nevada

On August 25, 1907, the *Washington Post* published an article entitled: "How Transient Sheep Business Was Ended in National Forests." The text analyzed how the nomadic pastoralism was "successfully" restricted in California's Sierra national forests and some adjacent areas of Nevada with the new administration of the United States Forest Service. The article praised the pioneering work of Chief Forester Gifford Pinchot and the new national forests for excluding large itinerant sheep bands from government forest lands. It attacked the sheep industry's transhumant or migratory use of grazing resources on the public domain and the foreign character of its Basque workers:

> In the old days of the Sierra Nevada Mountains, before any forest reserves or national forests as they are now called existed there, flourished an industry in Eastern California from Bakersfield to Lake Tahoe which was detrimental to home-building and permanent stock-raising. This was the immigrant or transient sheep business. . . . These sheep were driven in bands of from 2,000 to 3,000 each and were principally owned by Spanish and French Basques and Frenchmen who were mostly aliens and who owned no land and prevented paying taxes when possible. Some of the bands were owned by businessmen of Bakersfield who had Basque and French foremen to run them.[1]

The article noted that the new Forest Service policies were ending the transhumance grazing system in the mountains of California and Nevada, which was, according to the article, largely engaged in by Basque immigrants. It described Basque sheep owners of the "wandering bands" as mere *birds of passage* who intended to make some money and ultimately, after some years, return home.[2] But at the same time, this article recognized that some Basques "entered other business, some have bought ranches and are taking even chances with the American stockmen and have become good citizens of the United States."[3] Focusing on their contributions to the United States, this article reflects how at that time Basques were already integrating into the American society, although many others, in their status of nomad sheepherders, were seen as social parasites.

The *Washington Post*, however, paid little attention to the unregulated public-domain lands beyond the Sierra in the high desert of Nevada's Great Basin that were not regulated as national forest and where there were still large free and open rangelands. Despite the Forest Service policies, which restricted access to lands now deemed national forest, transhumance grazing by Basque and other sheepherders continued to be a widespread practice on these public-domain lands. In the nineteenth century, limited privatization of Nevada lands occurred and the largest private landowners were the Southern Pacific Railroad and a handful of prominent livestock companies. The rest of the lands within the boundaries of Nevada remained in the hands of the federal government. In fact, Nevada contained the largest percentage of unclaimed public-domain lands in the Far West, excepting possibly Alaska.

Despite the ongoing opportunities for grazing on public-domain lands, the government land policies, enacted as part of the broader conservation movement, would have a significant impact on the Basque sheepherders in Nevada, particularly nonlandowners.[4] At the turn of the century, Basque immigrant sheepherders, with their flocks coming onto public grazing lands, had been perceived as a serious threat to the cattle economic interests, sustainable agricultural development, and public interest in general. Basques were blamed for a number of interrelated economic and environmental problems affecting the western public grazing lands. Since many of the sheepherders in California and Nevada relied on the public-domain lands for their operations, the policies that the government enacted at this time would have an important impact on the sheep industry, including on the Basques who made up notable numbers of the sheepherders. Thereof, the new conservation movement that was gaining national influence at this time would turn its focus on the sheep and cattle grazing in the high mountain ranges in the West.

The early conservation measures drew a distinction between legitimate and illegitimate uses of natural resources. Far beyond natural resource governance, as historian Karl Jacoby has explained, conservation was a political and cultural process of state formation and consolidation. According to Jacoby, this process resulted in the creation of a bipolar narration in which conservationists were the "crusading heroes" and their adversaries the evil villains. Among those who were categorized as "small-minded, selfish villains" were the Basque sheepeherders.[5] Legitimacy—the most enduring issue in western history in Patricia Nelson Limerick's words[6]—would become Basque sheepherders and their communities' main pursuit in their long social integration process into mainstream society. This chapter analyses the socioeconomic impacts of government land-use regulation in some parts of Nevada's public domain upon the Basque sheepherders.

Nevada's Problems with the Free and Open Range

During the 1890s, Nevada's stock operators faced many difficulties and adjustments. The informal customs and rules of the past did not work. The "White Winter" revealed the fragile conditions of Nevada rangelands. Big stock operations had collapsed after their exploitation of the ranges and their monopolization of water resources to control the ranges. While smaller operations scrambled to pick up the pieces, the ranges were now open to outside herds, especially sheep. The new competitive situation caused new conflicts, especially on the already overgrazed ranges. At the turn of the century, an emerging conservation movement sought to bring order to the free and competitive use of some high mountain ranges in the West.[7]

Exploitation and degradation of the natural resources raised concerns among forestry and water experts as well as those Progressives who saw opportunities to champion the causes of an emerging conservation movement that emphasized "wise use" of resources based upon a utilitarian doctrine that stressed use for the greatest number, for the greatest good, for the longest period of time. In the 1890s, Congress initiated a partial reorganization of the public domain. In response to calls to conserve the nation's forest resources, Congress gave the president the authority to proclaim forest reserves on the public domain mostly in the Far and Mountain West. Establishment of forest reserves was a first step to conserve and protect conservation of water and timber resources.

With the passage of the Land Revision Act of 1891, Congress prohibited land entries or private land claims under the land laws of the United States on the designated or proclaimed forest reserve lands. The immediate effects of the Land Revision Act were to close resource use development on these lands. Historian William Rowley has contended: "The move toward regulated grazing in the forest reserves became part of an effort to achieve stability and permanency in an industry that had experienced destructive competition and resultant devastation of range forage."[8]

During the 1890s, several administrations set aside millions of acres of federal lands as forest reserves across the United States, particularly in the West. The federal government reserved forested lands to prevent their monopolization and destructive logging practices. The reserved lands also included grazing lands and their forage resources. The celebrated conservation movement took pride in the creation of forest reserves.

However, during the 1890s, no forest reserves were initially created in Nevada, a situation that sharply contrasted with its sister state, California. In California, the entrance of government management of renewable natural resources on public-domain lands altered significantly the economics of sheep production. The establishment of national parks and forest reserves in California closed access by sheep graziers to parts of the Sierra. It is well known that in 1891 Captain Abram E. Wood, commander of the Fourth Cavalry and first acting superintendent in charge of the troops assigned to protect the Yosemite National Park, drove out bands of sheep. Woolgrowers accused the forest reserves and later the national forests of permitting cattle to graze and excluding sheep. Typically, local groups, especially sheep and wool interests, rallied against grazing-use regulations in the new forest reserves.[9]

Early forest rangers were charged with the sometimes challenging task of preventing itinerant herders from entering government forest lands that were now administered under grazing regulations. Even earlier before the enforced regulations of the national forests, itinerant sheepherders wantonly disregarded any prohibition on their use of forest reserve grazing lands and developed contempt for any restrictions on the movement of their sheep bands. In the early twentieth century, Charles H. Shinn, an early forest ranger, recalled that sheepmen mocked and ridiculed the authority of the forest rangers. Oftentimes, sheepherders, Basques and others, violated the law and circumvented rangers' mandates.

In a popular novel from 1906 called *The Flock*, novelist Mary Austin, while describing sheepherders' experiences in California's Sierra Nevada, narrated as follows how a Basque sheepman often slipped past the inexperienced rangers:

> There was a Basque . . . a trick of his which served on more than one occasion was to start a small band moving, for he had fifteen thousand head, and having attracted the ranger's attention by boasts and threats made with the appearance of secrecy, in places most likely to reach the ranger's ear, to draw him on to following the decoy by suspicious behavior. Then the Basco would bring up the remainder of the flocks and whip into the Reserve behind the ranger's back. Once a day's journey deep in the Sierra fastnesses, it would be nearly impossible to come up with him until, perhaps, he neared the line on his fall returning.[10]

Basque sheepherders would also make appearances in other novels and popular literature of the time. Hamlin Garland in his *Cavanagh, Forest Ranger* (1910), a typical conservation novel of the era, described Basques as badmen and selfish villains: "It did not matter to him that these herders were poor Basques; it was the utter, horrifying, destructive disregard of law which raised such tumult in his blood."[11]

Despite this kind of popular representation of sheepherding, forest rangers like Charles Shinn complained about the cattle ranchers as well because of their opportunistic stance on the sheep question. Cattle people rushed to endorse the widely held opinions that roving bands of sheep destroyed forest vegetation, reduced water supply, and made for erosion.[12]

Some Basque sheep graziers were arrested and served jail time for violating government regulations. On July 21, 1893, the *California-ko Eskual Herria* [California's Basque Country], a weekly Basque language newspaper published in Los Angeles, informed its *euskaldun* readership that a Basque sheepherder, Piarres Ainchano, was arrested for trespassing on the Cucamonga forest reserve in San Bernardino County with his sheep band.[13] The next day, a Saturday morning, Ainchano paid the bail for release. Later the same day, Ainchano's employer Bertrand Elgart, who was also Basque, was also arrested for grazing sheep on the forest reserve. Elgart was released on the following Monday. The editor of the *California-ko Eskual Herria* blamed the government for arbitrarily imposing restrictions against grazing on the forest reserves. The newspaper noted that the boundaries of the new forest reserves were not widely known in the Basque grazing communities and offered to provide maps and information on the newly proclaimed reserves in order to prevent these incidents in the future.[14]

California sheepherders, however, were not the only ones taking issue with the newly instituted land regulations. Before long members of Congress of the far western states began demanding some answers on how the resources of these lands were to be utilized. Some accused the federal government of banning resource use entirely on the reserves, plotting to turn them into preserves or parks. In response, in 1897, Congress would pass the Forest Organic or Forest Use Act. The Act permitted resource use in the reserves, but only under the regulation and supervision of the Department of the Interior from where the original forest reserves were administered. By 1898 and 1899, the Department of the Interior and its General Land Office moved to regulate and limit the numbers of stock grazing on the newly established forest reserves. In March 1898, Nevada's *Reno Weekly Gazette and Stockman* informed its grazing community that the federal government opened the forest reserve in California "to her starving herds."[15]

While forest reserves were proclaimed in California during the 1890s, lands in Nevada would not be reserved for national forests until after 1905. Since there were no forest reserves or national forests in Nevada during the decade of the 1890s, the free and open range existed even in the valuable high-mountain pastures. In many respects it became a free-for-all as graziers competed for resources, sometimes violently. In the 1890s, the continued free and open range in Nevada invited sheepmen from California who no longer had access to the extensive forest reserves in that state. In 1894, famous preservationist John Muir described the invasion of Nevada in his *The Mountains of California*: "Immense numbers of starving sheep and cattle have been driven through them [the Sierra] into Nevada, trampling the wild gardens and meadows almost out of existence." The situation intensified the competition for forage and water resources on the Nevada ranges.[16]

Unregulated livestock grazing on common land invited crises and conflicts. In the still abundant, free open ranges of Nevada, law enforcement was weak or absent, giving opportunity to persistent incidence of violence. Confrontations occurred among landowning stock operators and nomadic sheep graziers over water rights, trespass, and destruction of forage resources on the common rangelands. Historian Samuel Hays wrote that "resorting to force and violence, sheepherders and cowboys 'solved' their disputes over grazing lands by slaughtering rival livestock and murdering rival stockmen."[17]

The growing anarchy on the Nevada ranges often made contracted sheepherders the brunt of the resulting range conflicts. Sheepherders, either Basques or others, were easy targets for their opponents or enemies. Their presence and especially their immigrant status created in part an agenda for conflict that marked the Nevada range struggles from the 1890s to the 1930s. In November 1895, two Basque sheepherders were killed at a sheep camp northwest of Pyramid Lake near Winnemucca Valley, in Nevada, by two ranch hands working for the cattle operation of J.M. Flannigan.[18]

By the turn of the century, problems on the Nevada ranges called for some authority to order the growing tensions between itinerants and landowning stock operators among both cattle and sheep owners. The unstable conditions created a destructive competition for the range resources. Among cattle people, however, there were often accepted agreements in the cattlemen associations with respect to branding of free-range cattle to denote ownership. But the new influx of sheep stood outside any type of orderly range-use system. Consequently, confrontations over range use and access to water resources were ongoing.[19] Inevitably, these disputes would involve many of the Basque sheepherders working throughout the region.

On January 15, 1901, Juan Bidart, a Basque sheep owner based in Battle Mountain, Lander County, "wrongfully and unlawfully" drove 5,000 head of sheep on the land leased by C.H. Duborg and Henry Anderson. On December 18, 1900, the Central Pacific Railway Company had granted to C.H. Duborg by a lease agreement lands in Lander County for the term of one year, beginning on January 1 and ending on December 31 of 1901. In turn he subleased to Henry Anderson half of his interest. From the first day of January 1901 and during the whole term mentioned above, then, Duborg and Anderson became the lessors of these parcels.

On February 13, 1901, C.H. Duborg's and Henry Anderson's attorneys lodged a complaint to the Third Judicial District Court of the State of Nevada in Lander County, that on January 15 Juan Bidart trespassed on the leased land of the plaintiffs.[20] They claimed damages for injury by the driving of sheep on these lands in the sum of $1,000. Furthermore, Duborg and Anderson asked for the attorney's fees and the court costs in the sum of $500. The plaintiffs, then, claimed $1,500 from Bidart. The judgment went against Bidart. On February 26, 1901, Summerfield wrote a letter from Reno to T. C. Malloy, county clerk of Lander County, notifying him that on February 24 both parties had signed a stipulation that "upon his [Juan Bidart] paying all of the costs of both the sheriff and yourself, the [trespassing] case may be dismissed."[21]

On the next day, February 25, the Judicial Court dismissed the lawsuit against Bidart upon the payment in full of the clerk's and sheriff's costs by the latter. Bidart paid up.[22] Nevertheless, the case *Duborg and Anderson v. Bidart* revealed the challenges of law enforcement regarding range matters. While Juan Bidart was being tried, Nevada's legislature considered a bill that would restrict itinerant sheepherders and protect property and small ranch owners within the state.

Introduction of National Forests in Nevada

In the early twentieth century, Nevada offered good opportunities for open-range stock grazing because of free range and rising agricultural prices. The surveyor general of Nevada, Edward D. Kelley, continued to be optimistic about the livestock industry in his biennial report for the years 1903 and 1904. He declared that the livestock industry was one of the most "important and profitable" economic sectors in Nevada, "and those engaged in it are among the wealthiest people in the State." The increasing arrival, however, of new stock outfits meant more competition for depleted ranges. Although the most productive lands were occupied and controlled by specific livestock and railroad companies, the remaining marginal rangelands still presented opportunities for the expansion of grazing enterprises. The county assessors' reports showed that the total estimated number of sheep population in Nevada was more than in the previous years.[23]

The scarcity of timbered lands in this arid state made it difficult for Nevada to qualify for forest reserve designations. On August 10, 1904, Gifford Pinchot, chief of the Division of Forestry, U.S. Department of Agriculture, and on behalf of the Public Land Commission, sent a letter to John Sparks, Nevada governor, asking some questions concerning lands under state ownership and their relation to the open-range public domain.[24] The responses that Kelley, the surveyor general, who also served as the state land register, sent from Carson City, generally indicated that few regulations were in place and much of the land in Nevada was marginal. Some of the questions and answers referred to were as follows:

> PINCHOT: What class of lands are most in demand?
> KELLEY: Agricultural and grazing lands are in greatest demand.
> PINCHOT: What protection is given to the State lands against fire and trespass?
> KELLEY: No protection except the general law against setting out fires, and a special law, which concerns individuals only, against trespass.
> PINCHOT: Is the State taking any steps to use and perpetuate the State timber lands, and if not, is special legislation required before such action can be taken?
> KELLEY: None of the reverted lands now owned by the State are timber lands.[25]

The responses reflected how at the start of the twentieth century, Nevada's public lands largely remained unregulated and unprotected. It would not be until after 1905 with the transfer of forest reserves from the Department of the Interior to the Department of Agriculture that national forests would be created in Nevada under the administration of the United States Forest Service.[26]

The administration of Theodore Roosevelt (1901–1909) invigorated the conservation movement. Gifford Pinchot persuaded the president to move the forest reserves to the Department of Agriculture under a new national forest system. In February 1905, the Department of the Interior's administrative functions and responsibilities were transferred to a newly established agency, the United States Forest Service, within the Department of Agriculture. Gifford Pinchot, friend and adviser of Roosevelt, was appointed chief forester for the new agency. At the turn of the century, to Pinchot, grazing was one of the most important economic activities in the forest reserves. In 1905 Pinchot and the Forest Service continued to implement a permit system for grazing that the General Land Office had begun. The Forest Service issued a *Use Book* that defined resource-use regulations under a permit for an allotment. In 1902, the Department of the Interior through the *Forest Reserve Manual* had stated that the secretary of the interior possessed the authority to restrict any livestock grazing activity with the aim of protecting the forest reserves. The new Forest Service blueprint made similar claims for the secretary of agriculture.[27]

The *Use Book* set out rules on grazing seasons, set numbers of stock, and issued grazing permits based upon property ownership and traditional use criteria. Three classes of grazing permits were available: first, Class A for those owners of ranch properties within or adjacent to the national forests who customarily grazed stock on lands now within the national forests; second, Class B for those who possessed property near the national forests and traditionally grazed stock in these high-mountain pastures; and third, Class C for itinerant graziers who did not own property or a home ranch. The various permits were issued on the basis of this preference criteria. All permits granted were considered grazing privileges, not rights, by the Forest Service. The preference system favored the first two classes and generally excluded the third or Class C permits. Also, as William Rowley has explained, the Forest Service based its preference system on the concept of "commensurate property ownership." The principle of commensurability required graziers to own enough private land to support winter feeding of stock when they must be removed from the national forests at the end of the season of grazing. Forest Service considered these policies the mechanism to defend established landowning ranchers and small homesteads against the intrusion of itinerant sheep graziers, typically linked with Basques.[28]

Unconventional Conditions: The Report of Herbert Stabler of 1906

From the very beginning, itinerant sheep herding activities, in which Basques were largely involved, shaped the early decisions of the U.S. Forest Administration in Nevada. Just as itinerant sheep raising was increasingly associated with a Basque immigrant culture, Basque sheepherders were potential and frequent targets in the justification of the expanded government role in Nevada's public-domain lands.

Forest inspectors went far and wide into Nevada to mark off possible forest reservations. New national forests in the high mountain ranges meant the extension of grazing regulations in the higher and better watered elevations. From April 1 to May 1, 1906, Herbert O. Stabler, forest assistant, prepared a detailed report based on an examination in the Monitor Range in central Nevada. The title of the report itself anticipated the establishment of a new forest: "A Favorable Report on the

Proposed Monitor Forest Reserve, Nevada." It noted forest resources in this part of central Nevada, especially the protection of vegetation, especially ground cover, to protect water supplies. The report was addressed to the chief forester Pinchot, which would serve as a recommendation to President Roosevelt to create a national forest by proclamation under the authority granted by the provisions of the 1891 Land Revision Act.

The proposed Monitor Forest Reserve was located in Eureka and Nye Counties. While the northern boundary was marked fifteen miles north of the Nye-Eureka county line, the southern boundary was about seventy-seven miles south of the Nye-Eureka county line, nearly due east from Tonopah. The proposed reserve comprised roughly 450,000 acres. Only one-fourth to one-third of the area was surveyed. During the nineteenth century and later, most of Nevada was not surveyed into the rectilinear townships and ranges required in the Land Ordinance of 1785. The Monitor Range was one of a number of parallel mountain ranges that extend north to south in central Nevada. The valley basins between these mountain ranges were at high elevations between 5,000 and 6,500 feet. According to Stabler, the scarcity of water resources prevailed, but there was an exception in the Reese River Valley, where there was, in his own words, "a continuous stream of water." He also noted the persistence of a severe drought that affected and altered dramatically the ecological and agricultural situation. The condition accelerated the process of range degradation in Nevada's marginal lands: "In the other valleys the creeks that flow into them are soon dried up and lost because the soil is very sandy and the gradient slight." In terms of climatic concerns, Stabler stressed that these valleys were characterized by the scarcity of rains and very low temperatures, with considerable snowfalls. He observed that the extremely cold winter temperatures often caused great losses of cattle to the ranchers.[29]

The proposed reserve consisted mostly of piñon and juniper trees, which were located primarily on the base of the mountains. Nonetheless, Stabler noted, the distribution of the forestland tracts was very irregular. There was, in his estimate, "25 percent to 40 percent of the area shown as woodland that has no forest on it at all. In places where the stand is good it will yield 4 to 5 cords per acre, but as a whole the type will not average more than 1-1/2 cords per acre. Limber pine is found occasionally but in an inconsiderable quantity." The previous mine development in the Danville Canyon in the southern part of the Monitor range had displayed the problems regarding the scarcity of lumber supply. According to Stabler, nonetheless, a scattering reproduction constantly took place that was enough to replace the older trees. In that respect, he noted that "if an area is out clear [cleared out or cut over], or nearly so, a long period is required for the development of a normal stand."[30]

Stabler explained the importance of water and timber conservation to assure the protection of stock ranching as the main economic activity. Looking to the future, Stabler believed that new mining booms would promote local population growth. The protection of forest resources, especially water supplies, was necessary for future economic development: "In that case something could be done toward storing water in some of the canyons and in this way many acres could be irrigated and the towns would furnish a ready market for all products." He added: "In no event, however, would a protective forest be a serious consideration [there was no forest], because the greater part of the area really [was] valuable for holding water [possible reservoir sites]. . . ." In absence of significant forests to protect water resources, the forest inspector was saying that these lands should be reserved because reservoirs could be constructed on these lands to ensure adequate water supplies.[31]

The report assigned considerable importance to the prevailing economic activities on those lands, which were under consideration for the creation of a new national forest. In the Monitor range, the forest assistant wrote, "there is not alienated agricultural land within the limits of the proposed reserve." Stabler did locate three ranches—two devoted to cattle raising and the other one to sheep—on the western part of the range at the mouth of the creek and outside of the proposed forest. Each of these operations had ten to fifteen acres of irrigated land for hay production and other grassland to secure livestock feed during the winter. In addition, there were other cattle ranches on the west side of the Monitor Valley and on the east side, two other operations devoted to cattle raising as well. While the livestock ranching did not usually depend on a supply of timber, according to the Stabler's report, there were other enterprises in the Monitor range that did.

For instance, on the western part of the proposed reserve, there was a pumping plant that supplied the town of Tonopah with water. The plant required about 1,300 cords of wood per year for its pumps. Also, a mining company located at Hannapah also depended on a wood supply, but the introduction of coal and electricity helped reduce cordwood consumption. Stabler also noted that Hannapah was a declining settlement and the existence of a forest reservation would now help the development of new communities.

The report paid special attention to the livestock grazing activities in the vicinity. Stabler estimated between eight hundred to one thousand cattle and at least 88 percent of this livestock was owned in the Monitor Valley. On the other hand, he mentioned the case of the Potts sheep ranch, which operated five thousand head of sheep separated into three bands. Stabler observed that the Monitor range was better suited to sheep than cattle. In recent years, he said, the sheep population had increased in this part of the country because sheep generally grazed this mountainous range more efficiently than cattle did. Furthermore, the profitability of sheep supplying the new mining towns in Nevada and a favorable international market had increased demand for sheep products. He observed many cattlemen were considering converting their herds to sheep, although the latest changes in the market were favorable to beef. Stabler explained this process as follows:

> The range is better suited for sheep as they do not require as much water as cattle, and can, and do graze in rough, rugged portions of the mountains that cattle can not get to. Sheep have been very profitable of recent years owing to the fact that mining camps in the Toquima range [Nye County, Nevada] have furnished a near-by market for mutton and the drought in Australia has caused wool in this country to sell at a good price. Ranchers, now ranging nothing but cattle, are seriously considering selling out their cattle and buying sheep. Such a step has been postponed because of the good market now available in the mining camps.[32]

According to this forest assistant, overgrazing in the valleys and in the mountains was adversely affecting the livestock agriculture. In his words, grazing was "very poor" on the Monitor range with the exception of some extensive areas of white sage or "winterfat" that made for nutritious grazing, especially in winter.[33] In addition, he made reference to the issues between cattle and sheep ranching in this country. Apparently, to Stabler, there existed a harmony among the local cattle and sheep breeders. The problem was outsiders: "The transient sheep question causes hard feeling and is a matter for immediate regulation if a reserve is created."[34] He further noted briefly that the danger of fire in that part of the country was at a minimum level and in case a fire started, he wrote, "it rarely spreads because there is no ground cover for the fire to run in."[35] With the growing threat of transient sheep invasions, public sentiment in the community, according to Stabler, was "strongly in favor of forest reserves." As he explained, local stockmen saw the creation of a national forest as a means to protect their own economic interests from itinerant sheep herds. In Stabler's opinion, the creation of a forest reserve would provide future mining towns with timber of all kinds. At the end of the report, considering all the factors, Stabler recommended the creation of a national forest in the Monitor range primarily because it would reinforce economic development and enhance the settlement in central Nevada, and furthermore, remove "many transient sheep" from the range.[36]

Stabler's forest reserve proposal's report argued the economic importance of setting aside tracts of public lands for forest conservation, while acknowledging that the Nevada environment did not present conventional conditions for the creation of a national forest. Local livestock operators also complained that the government allowed wild horses to compete with the domestic livestock during drought years. But also, and perhaps more significantly, they blamed the transient stock operations and their Basque immigrant herders for scavenging the scarce forage.

For the Forest Service, it was hard to justify the costs of administration to protect the sparse resources in the name of conservation in Nevada. The lands appeared desolate and degraded, but one of the reasons they were in such a condition was because of the damage that sheepherders were inflicting. Around this time, Basques, who were widely associated with the itinerant sheepherding in these ranges, and whose "foreignness" would start to be called out by pro-cattle interests in the ensuing policy debates, would become a scapegoat, as Kevin Hatfield has contended, "for the ecological and social ills of the public range."[37]

Mark Woodruff and the National Forests in Nevada

By the 1900s, there were a number of Basque sheepmen who had established properties in the areas close to or within the rural areas now declared national forests. Early Forest Service reports testified how by the 1900s it was an energetic Basque immigrant community owning land and prospering in the open-range sheep industry. The Basques, evidently, had come to stay. However, the image of the Basque immigrant sheepherder as landless and nomadic continued to live on in the public imagination for years, and the ethnic association of Basques with itinerant sheepherding would be a lasting one.

On November 20, 1907, Mark G. Woodruff, forest supervisor in Nevada since July 1, sent from Austin (Nevada) to Washington, D.C., a detailed report about the management of the Toiyabe, Toquima, and Monitor national forests. Woodruff made a detailed investigation and gathered data relative to the range conditions in the protected lands. In this grazing report, Woodruff explained in detail the range forage degradation on the forest ranges as a consequence of overgrazing and land abuses. It was a distressing situation: "The Toiyabe range has been badly overgrazed as a whole; the Shoshone range is in fairly good condition; the Desatoya range supplies such a small amount of grazing that it is of no particular consequence; the Toquima range has been abused along the foothills, and the Monitor range has been overgrazed in parts."[38] To understand the nature of the problem, the report provided first some relevant background information, which summarizes the local grazing development stemming from the prolonged crisis since the winter of 1889–1890, with special focus on the expansion of sheep grazing to the detriment of cattle ranching: "The sheep men and their flocks have become more numerous, and finally disregarding any and all rights belonging to other range users, the sheep men have encroached even to the fences of the patented lands in the low foot hills." He further noted:

> The plan of grazing has been to winter the sheep in the deserts lying to the south of these Forests. As spring comes on the sheep work north to the country lying about thirty miles north of Austin, where the lambing and shearing takes place. As soon as the lambs can travel they are started south again along the top of the Toiyabe range. The best feeding ground is at the head of the north fork of the Reese river and Twin canyons. Every flock owner has tried to reach that country first. It is a race from the start. The result has been to literally cut the surface of the Toiyabe range into the sheep trails, the bottoms of which in places are several inches below the roots of the grass and weeds.[39]

Woodruff made the expansion of sheep grazing synonymous with the range problem. The forester noted that the sheep industry had become the leading agricultural economic activity in central Nevada

Table 3.1. Sheep companies in Central Nevada

Company	Number of Sheep
The Lander County Live Stock Co.	22,000
Mrs. George Watt	4,000
John Spencer & Co.	8,000
J. A. Miller	7,000
John Laxqua & Pete Lardapeda	5,000
Borda & Co.	5,000
Pedro Equira	4,000
Isadore Sara	8,000
Those turned back in 1907 were:	
Espiel Brothers	3,000
John Echevara	5,000
Esparand	5,400

Source: Mark G. Woodruff to The Forester, Washington, D.C., November 20, 1907, Record Group 95, U.S. Forest Service, Files Office of the Chief, Grazing Plans, National Records and Archives Administration, College Park, Maryland.

threatening cattle ranching. Woodruff reported how sheep "cut the gravelly . . . soil very badly" and noted his efforts to protect ranges against the ravages of thousands of sheep: "You will realize more fully what the condition is when I state that in 1906 ninety-six thousand sheep made the trip from the north end to the south end and back to the starting point through the narrow Toiyabe range." He said that he succeeded in stopping 13,400 sheep of a total 63,000 from entering the national forest. He made an estimate of the total sheep population crossing the national forest counting the livestock of the operations based primarily on small operations.[40] As it is shown in table 3.1, roughly half of these companies were owned by Basques.

Roving sheep bands on Nevada's open ranges, according to Woodruff, were a grave threat and he considered it of vital importance to stop the expansion of these itinerant sheepherders. Woodruff often took action against sheepmen by ordering them away from water sources used by locally owned cattle interests and giving them only a Class C permit that usually denied them any opportunity for a grazing permit on the national forests.[41]

The report continued analyzing the situation in the Monitor range, where again the greatest problem that was threatening the local economy, the report concluded, was the sheep industry. Particularly, the east part of the Monitor range, Woodruff contended, was highly damaged by the sheep of the Basque-owned Eureka Livestock Company: "The district so affected runs from the north line to the south line of Township 15 N. These hills have been used for both sheep and cattle."[42]

Despite his negative view about the increasing numbers of sheep in the national forest lands, Woodruff recognized that the good economic conditions for sheep products gave sheep graziers impressive economic returns on the sheep admitted to the forest. While sheep became an issue during the implementation of early forest management, the same held true for wild horses that roamed free and shared pastures with livestock. Along with the poisonous plants and predators, wild horses were main factors that affected greatly the sheep industry. Wild horses were overrunning the national forest lands of Nevada and by far were the "greatest destroyers of range" Woodruff said, because they trampled the best pastures and water resources and thus threatened the forest ecosystem. He considered wild horses "a pest, a nuisance, inbred, worthless and [that] annually destroy forage that would sustain at least 2,500 cattle." Woodruff wanted to remove every wild horse from the ranges of the national forest.[43]

After analyzing the factors bearing upon cattle livestock grazing, Woodruff considered it necessary to divide the range between sheep and cattle operations. By applying standard scientific criteria (estimating range potential), he intended to reorganize the range in favor of cattle, offering the sheep graziers the most rugged terrain since their stock could utilize the more difficult ranges. To put it another way, he wanted to remake "an ideal cattle range." Woodruff's plan was to keep both classes of stock separate "by flagging the lines," assigning the best pastures for cattle and keeping sheep in the more inaccessible marginal pastures where sheep could be the more successful grazers. Furthermore, he urged his superior to "expend considerable sum of money" in the rehabilitation and restoration of water resources to support the varied stock operations.[44]

The land-use reclassification had to meet the general plan for managing grazing resources in the national forest according to the *Use Book* and the classification of grazing permits. The main goal was always to, in Woodruff's words, "correct range abuses and to give the degree of protection to those people who are permanent residents, land owners, and of value to the state, that is required in order to build up a prosperous community." He noted that the Forest Service should take "a positive position in favor of the home builder as against the transient herder." The conservation of these pastures had to consider only, in his opinion, the interests of the "large land owners near the national forest, who are investing in more lands year by year, who own substantial homes and who are adding to the development of this state."[45]

Woodruff explicitly criticized the Basques who operated in central Nevada. He related the career of two Basque sheepmen Isadore Sara and John Borda, to whom the forester gave a Class C permit:

> I find that he [Isadore Sara] first came into the country as a sheep herder for a man named Jensen. That was about 1893. In 1903 he acquired an interest in a band of sheep formerly owned by Dixon & Stebbins. Then he sold out to George Watt. He then moved to Tonopah, returning here in 1905 and again entering the sheep business. He owns no land, does not own the house in which he lives, and has nothing in common with the residents of this vicinity. He is a Basque, but has taken out his first citizenship papers. I have classed him as Class C.[46]

He had a similarly low opinion of Borda:

> All that has been said for Sara can be said of John Borda, except that he has used the range continuously. He owns nothing aside from sheep and I have put him in Class C in my list of recommendations.[47]

Woodruff saw these people deserving a Class C permit because they undermined respect for the law, order, and private property. He asserted: "Another important reason for so classing men who own no land is that the complaint comes from all directions from the cattle and sheep men alike, that it is the Basque sheep owner and herder who commits the greatest damage to private interests and who utterly refuses to recognize the rights of cattle owners to any range." He emphasized how cattlemen did agree to share the range with sheepmen landowners, but not with the transient operators: "The cattle men and ranchmen do not seriously object to dividing the range with local sheep men who own land, but they seriously object to allowing transient flock owners to acquire range rights, or to a recognition of acquired rights, secured by forcing themselves upon the range during recent years." Despite the fact that there were already some Basque sheep operators who owned land, this immigrant community continued to be associated with sheep itinerancy. Woodruff observed a broad objection against transients in the local community and welcomed more regulations that could eliminate them. Indeed, the establishment of national forests in Nevada served to guarantee the economic interests of those who owned land: "Local residents welcome Forest administration solely upon the ground that they need protection in range matters. They feel that they will be driven out of the country unless it is given, and as a matter of fact there is very little reason for the existence of national forests here aside from water-shed and grazing protection. I strongly believe that the Basques should give way to the man who owns something." Considering the particular case of Nevada and to avoid tramping the cattle range, he suggested that sheep and cattle be kept in different grazing districts, and he further recommended "that all sheep owners be notified that upon over-running their boundaries a second offense will forfeit their permits."[48] The social stigma carried by the sheepherding job and activities that arose around this time would become detrimental for the Basque immigrant community linked to this industry.

At the time of the creation of the national forest, there was no livestock association in this district. Woodruff himself pressured both local cattle and sheep operators to organize themselves and do everything possible in their power to assist administration to protect the range. Livestock operators appeared enthusiastic about the idea of setting up an association as a means to "provide a permanent and reliable grazing ground in future years." Woodruff judged important the establishment of a livestock association because it could serve effectively to mediate misunderstandings between the Forest Service and local graziers over *Use Book* regulations, and also "settling range disputes that may arise" among private parties.[49]

Woodruff prepared a grazing map showing the range divisions within the Nevada national forests. He proposed the following livestock restrictions depending on each rangeland's conditions for grazing in the year 1908 on three Forests:

Table 3.2. Livestock restrictions in the national forests of Nevada

Livestock	Toiyabe national forest	Toquima national forest	Monitor national forest
Cattle	8,000	5,000	2,000
Sheep	28,500	No sheep	10,000
	Shoshone range		
Cattle	1,500		
Sheep	5,000		

Source: Mark G. Woodruff to The Forester, Washington, D.C., November 20, 1907, Record Group 95, U.S. Forest Service, Files Office of the Chief, Grazing Plans, National Records and Archives Administration, College Park, Maryland.

For the case of the Toquima National Forest, Woodruff prohibited the entrance of sheep altogether and established a forest guard to keep them out. He made a point to notify every sheepman of these instructions. Many sheepmen used the Toquima range to secure water, while during the fall they went south to the desert and returned north in the spring.

While reducing the number of sheep in these forest ranges, Woodruff proposed to grant the permits, Classes A and B, to the following stockmen for usage rights to these ranges:

Table 3.3. Sheep grazing permits holders

Company	Number of Sheep for 1907	Number of Sheep for 1908	Class
The Lander County Live Stock Co., Austin	22,000	16,000	B
Mrs. George Watt, Austin	4,000	3,000	B
John Spencer & Co., Austin, sold 2,000 since close of season 1907	8,000	5,000	A
J. A. Miller, Austin	7,000	4,500	B
Total	41,000	28,500	

Source: Mark G. Woodruff to The Forester, Washington, D.C., November 20, 1907, Record Group 95, U.S. Forest Service, Files Office of the Chief, Grazing Plans, National Records and Archives Administration, College Park, Maryland.

In the lower level, or better said, in the Class C, Woodruff recommended the following transient livestock operators to be "excluded" from the national forest for the year 1908 because they did not own any land: John W. Freeman Company, John P. Williams, John Laxqua, Pete Lardapeda, John Borda, Pedro Equira, and Isadore Sara. He made clear that nonowners of land were not welcome:

- John W. Freeman Co., Fallon Nevada, sheep, owns nothing in or near any of these national forests and resides over one hundred miles from these ranges.
- John P. Williams, Fallon Nevada, and members of his family associated with him in said business, owns nothing in or near any of these national forests and resides about 100 miles distant.
- John Laxqua and Pete Lardapeda, Basques, 5,000 sheep on Toiyabe in 1907, own no land and are not American citizens.
- John Borda, Basque, 5,000 sheep, owns no land. Used forest a number of years.
- Pedro Equira, Basque, owns 4,200 sheep, owns no land, swore under oath that he was not an American citizen. Was allowed in the forest in 1907 on account of weak lambs.
- Isadore Sara, Basque, owns no land in or near national forest, sold sheep in 1907. Will purchase more sheep and attempt to range them on Toiyabe in 1908, owns no land or home. Has grazed there a number of years.[50]

In the Monitor forest, the range accommodated both cattle and sheep. In District 7, he explained that the best right to the sheep range belonged to Mrs. Morrison, who lived near the national forest on Allison Creek. In the same district, the Basque-owned Eureka Livestock Company customarily grazed their sheep there too. Thus, the forester proposed to admit there two bands of sheep counting 2,500 head each, owned by both the Morrisons and the Eureka Livestock Company. The Eureka Livestock Company had used these pastures for a long time already and in recent years expanded its range use. Since the Eureka Livestock Company was powerful enough to defend its customary rights, especially with their attorney, the influential congressman George A. Barlett (1869–1951), Woodruff proposed to let this Basque-owned sheep company to continue operating on the now national forest lands. Indeed, the Eureka Livestock Company was the only Basque operation with a Class B permit. Briefly, Woodruff's grazing management report proposed to restore the ranges in the national forests of Nevada by reducing the livestock grazing, favoring cattle, and reducing sheep by rejecting Class C permit holders. As a result, this report would have particular effect on the Basque sheepherders, since all the Basques, except the Eureka Livestock Company, possessed Class C permits.

At the end of the report, Woodruff recommended the sheep grazing season to start on July 1 and end on October 15. He stipulated the grazing fee for all sheep over six months of age at eight cents per head. Woodruff expected opposition from ranchmen to the prices, but he thought that those prices were affordable and he further thought that "it will be better to begin grazing administration by charging good prices than to commence at a low price and be compelled to raise it at a later date." He concluded, however, that the "feeling expressed toward national forest Administration has been almost unanimously cordial. Several people announce that if we can better conditions, they will welcome it, and in one or two cases they openly declare that they are opposed to it."[51]

Nonetheless, the beginnings of grazing administration raised constitutional matters. In 1906, the enforcement of the Forest Service policies and particularly the implementation of grazing fees sparked criticism from the livestock community. While most welcomed the new rules ordering the rangelands and excluding the nonlandowning sheepherders, they did not like the imposed grazing fees that the Forest Service said was necessary to pay for its regulatory administration.[52]

Not surprisingly a court challenge against the fee system was not far away. The Eureka Livestock Company took these objections into court. It filed a complaint against: Theodore Roosevelt, president of the United States; James Wilson, secretary of agriculture; Gifford Pinchot, first chief of the Forest Service; and David Barnett, forest supervisor in Nevada. The 1908 court case became *Eureka Livestock Company v. United States*. The Basque sheepman's complaint challenged the constitutionality of the federal government to set aside timber lands as national forests under an act of Congress in the Monitor range, in central Nevada. In September 1908, the plaintiff argued that no federal agency could set apart the public domain for conservation purposes and restrict access to grazing lands where graziers traditionally operated. According to the Eureka Livestock Company, when the Department of Agriculture established the national forest in the Monitor Range, the company was grazing about five thousand sheep divided into two bands. Under the new regulations, the Forest Service denied the livestock company access to these lands and threatened to confiscate the sheep if the company persisted in grazing there. The plaintiff claimed that these lands were only valuable for agricultural and mineral purposes and had little potential for timber extraction, something which made them unsuitable for conservation and, therefore, the claimant said, should not be included in the national forest. This was among several cases challenging the legitimacy not only of imposing grazing fees, but also the creation of the national forests themselves. In general, national forests in Nevada were established by both presidential proclamations and congressional acts, although in 1908, Congress moved to put the creation of new national forests under its authority denying the former power of proclamation to the president. Ultimately, by 1911, with the *Light* and *Grimaud* cases the United States Supreme Court ruled in favor of the legitimacy of the national forests and their authority to impose grazing fees.[53]

With the arrival of the Forest Service grazing regulations, order finally started to take hold in the high mountain pastures of Nevada's mountain ranges, an order that came at the expense of Basque sheepherding. Beyond the boundaries of the national forests, however, Nevada's public-domain ranges remained unregulated and a persistent grazing problem, where conflicts continued including Native American competition for range lands.[54] In 1911, a small band of Indians murdered four sheepherders, including Basques. This incident, commonly known as "the Last Indian Uprising," illustrated the still violent character of the American West.[55] Early forest inspectors looking to designate public-domain lands as national forests noted that Nevada was different. Compared with other states, including the western ones, the state of Nevada presented a challenge for the definition and justification of proclaiming national forest lands. As the free-range stock industry kept growing at impressive levels, the federal government continued a *laissez-faire* policy toward grazing on the public domain outside the established national forests wherein it did apply grazing regulations that often favored cattle interests.

The continuing crises in Nevada livestock ranching revealed a growing national debate over the future of the still unregulated and free and open-range grazing on the public domain. During the first three decades of the century, this situation increasingly pushed the Nevada legislature to become involved in range issues. Local experiments in range matters preceded the organization of grazing districts out of the public domain after the Taylor Grazing Act of 1934.[56] The narrative that followed all these events continued to center on heroes, embodied by the conservationists and the more venerable cattle graziers, and the villains sheepherders, which continued to be associated with a Basque immigrant culture in Nevada.

Summary

In 1905, with the coming federal land management to portions of the public domain in Nevada (mountain pastures), the Forest Service tried to limit or exclude large itinerant sheepherding from the lands it administered. Throughout the still-abundant public grazing lands beyond the Forest Service's national forests within the state, however, the itinerant sheep herds still moved into the basin and rangelands each season, and some overwintered on white sage or winter fat. While cattle grazing interests were generally pleased with Forest Service grazing policies, they were displeased with the lack of attention to grazing regulations and indifference to the chaotic grazing situation on the open rangelands of Nevada beyond the Forest Service boundaries. As a result, expanding Forest Service range regulations to these public-domain grazing lands outside the national forests was seen by many as the solution to halting sheep itinerancy and offering the prospect of more stability on the range.

This confrontation over land use and land rights would lead to a situation where itinerant sheepherders on the public-domain lands became widely denigrated, and would also involve the creation of the most pernicious image of Basque immigrants in the West as being the antithesis of the nationally venerated image of cowboys. While cowboys were viewed as the embodiment of the American frontier spirit through self-reliance and hard work, and who rode horses, sheepherders were considered despicable and of an inferior class of laborers who walked long distances allowing sheep to roam at will across the wide-open range.

The next chapter analyses how—at the same time as the Basques began to become settled in the Interior West with many Basques building prosperous sheep-dealing businesses—an anti-Basque movement began among cattle ranchers, conservationists, and some politicians. Eventually, a positive image of Basque immigrants and Basque sheepherding communities would reemerge in the decades to come, but not before they first became the target of wide-ranging political and social ire.

Notes

1. "How Transient Sheep Business Was Ended in National Forests," *Washington Post*, August 25, 1907, 8.
2. In the late nineteenth and early twentieth centuries, the expression "birds of passage" was coined to refer to immigrants coming with the intention of returning home after some years of work. Peter Roberts, *The New Immigration: A Study*

of the Industrial and Social Life of Southeastern Europeans in America* (New York: Macmillan, 1913), 13.

3. *Washington Post,* August 25, 1907, 8.
4. Clawson, "Range Lands of Northeastern Nevada," 4–6; Adams, "Public Range Lands," 324.
5. Karl Jacoby, *Crimes against Nature: Squatters, Poachers, Thieves, and the Hidden History of American Conservation* (Berkeley: University of California Press, 2001), 3, 6.
6. Limerick, *Legacy of Conquest,* 254.
7. Young and Sparks, *Cattle in the Cold Desert,* 219–27; Rowley, *U.S. Forest Service and Rangelands,* 15–21; also see for an overview of Progressivism's reform agenda, Hays, *Conservation and the Gospel of Efficiency;* Robert H. Wiebe, *The Search for Order, 1877–1920* (New York: Hill and Wang, 1967).
8. Rowley, *U.S. Forest Service Grazing and Rangelands,* 4, 5.
9. Harvey Meyerson, *Nature's Army: When Soldiers Fought for Yosemite* (Lawrence: University Press of Kansas, 2001), 95–103; "Conquering the West: The Federal Lands Question," *National Wool Grower* 55 (January 1965): 87–89; Rowley, *U.S. Forest Service Grazing and Rangelands,* 28–38; see also Lawrence Rakestraw, "Sheep Grazing in the Cascade Range: John Minto vs. John Muir," *Pacific Historical Review* 27, no. 4 (November 1958): 371–82.
10. Austin, *The Flock,* 197–98.
11. Hamlin Garland, *Cavanagh, Forest Ranger: A Romance of the Mountain West* (New York: Harper and Brothers, 1910), 183.
12. Jeronima Echeverria, "Basque 'Tramp Herders' on Forbidden Ground: Early Grazing Controversies in California's National Reserves," *Locus* 4, no. 1 (Fall 1991): 41–58; Charles H. Shinn, "Work in a National Forest," *Forestry and Irrigation* 13 (November 1907): 590–97.
13. *California-ko Eskual Herria* was a weekly newspaper in Basque language published in the city of Los Angeles, California, between 1893 and 1898.
14. "Artzainak eta Gobernamendua," *California-ko Eskual Herria,* July 22, 1893, 2.
15. Harold K. Steen, *The U.S. Forest Service: A History* (Seattle: University of Washington Press, 1976), 26–46; Char Miller, *Gifford Pinchot and the Making of Modern Environmentalism* (Washington, D.C.: Island Press, 2001), 119–44; Rowley, *U.S. Forest Service Grazing and Rangelands,* 22–31; "Relief for California," *Reno Weekly Gazette and Stockman,* March 31, 1898, 2.
16. Clawson, "Range Lands of Northeastern Nevada," 4–6; John Muir, *The Mountains of California* (New York: The Century Co., 1907), 96.
17. Hays, *Conservation and the Gospel of Efficiency,* 49–50. See also: Bill O'Neal, *Cattlemen vs. Sheepherders: Five Decades of Violence in the West, 1880–1920* (Austin, Tex.: Eakin Press, 1989).
18. Lane, *Anglo-American Contributions to Basque Studies,* 35–36; Young and Sparks, *Cattle in the Cold Desert,* 219–30; "A Double Murder Reported," *Nevada State Journal,* November 16, 1895; in February 1896, Jack Davis, better known as "Diamondfield" Jack, a hired gunslinger for the Sparks-Harrell cattle company,

allegedly killed two sheepherders in the state of Idaho close to the Nevada border. The "Diamondfield" Jack murder case, which has stirred interest among western historians, became a symbol of the Nevada–Idaho court history, see David H. Grover, *Diamondfield Jack: A Study in Frontier Justice* (Norman: University of Oklahoma Press, 1986), 7–44.

19. Young and Sparks, *Cattle in the Cold Desert*, 225; Hays, *Conservation and the Gospel of Efficiency*, 49–60; Clawson, "Range Lands of Northeastern Nevada," 6.
20. From 1863 until 1979, Austin served as county seat of Lander County, Dean Heller, Secretary of State, *Political History of Nevada (Eleventh Edition)* (Carson City: State Printing Office, 2006), 122.
21. *C.H. Duborg and Henry Anderson v. Juan Bidart*, case no. 2069, February 16, 1901; James D. Torreyson and Sardis Summerfield, Attorneys for Plaintiff; Sheriff's return, February 23, 1901, State of Nevada, County of Lander; Third Judicial District Court judgment notification to Juan Bidart. February 13, 1901; Sardis Summerfield to T.C. Malloy, Reno, Nevada, February 26, 1901, Trial Records, State Court, Lander County Courthouse, Battle Mountain, Nevada.
22. Ibid.
23. *Biennial Report of the Surveyor-General and State Land Register, 1903–1904* (Carson City: State Printing Office, 1905), 9.
24. In 1903, President Theodore Roosevelt appointed a public land commission composed of Gifford Pinchot, Frederick H. Newell, and William A. Richards in an attempt to determine the problems affecting the public-domain lands, *Report of the Public Lands Commission, With Appendix*, 58th Cong., 3d Sess., Doc. no. 189 (Washington, D.C.: Government Printing Office, 1905), 3–11.
25. Edward D. Kelley, State Land Register, to Gifford Pinchot, August 31, 1904, *Biennial Report of the Surveyor-General and State Land Register, 1903–1904*, "Questions and Answers," Gifford Pinchot, Secretary of the Commission on the Public Lands, to John Sparks, Governor of Nevada, August 10, 1904, 15–16.
26. Lane, "The Cultural Ecology of Sheep Nomadism," 142.
27. Miller, *Gifford Pinchot and the Making of Modern Environmentalism*, 162–64; Rowley, *U.S. Forest Service Grazing*, 46–59; United States Department of Agriculture, 1905 "Use Book": The Use of the National Forest Reserves; Regulations and Instructions (Washington, D.C.: Government Printing Office, 1905).
28. William D. Rowley, "Privilege vs. Right: Livestock Grazing in U.S. Government Forests," in *History of Sustained-Yield Forestry: A Symposium*, ed. Harold K. Steen (Durham: Forest History Society, 1984), 61–67; Rowley, *U.S. Forest Service Grazing*, 58–62.
29. Herbert O. Stabler, "A Favorable Report on the Proposed Monitor Forest Reserve Nevada 1906," 2–3, Folder: LP-Boundaries, General, Toiyabe National Forest, 1906–1941, Record Group 95, Federal Records Center, National Archives, San Bruno, California.
30. Ibid., 4.
31. Ibid., 5.
32. Ibid., 8.

33. See "winterfat" in Young and Sparks, *Cattle in the Cold Desert*, 61–62. In March 1896, John W. Freeman, a Nevada sheep operator, had written the following words to the *American Wool and Cotton Reporter*: "Nevada's soils are like her feeds, various brush, foliage, bunch grass, white and black sage are all found in the immediate vicinities. She has miles of such domain, and there are today over one-half million sheep roving over her 'hills and dales…' The mountain ranges produce bunch grass and sunflowers and are well watered. Upon these the sheep are run during the spring, summer and autumn. Upon her vast deserts a very rich feed known as 'white sage,' grows to a height of one or two feet. This feed is utilized when the 'snow is on the ground.' There being no water on the deserts the feed grows undisturbed during the summer; the stock owners are not obliged to put up hay as the 'bush,' sand grass and white sage produce such a variety of feed that 'Mary's little lambs' would leave hay stacks and roam at will. The herders laugh and grow fat when the beautiful snow falls and they fall down out of the mountains on to these deserts where there is such abundance of feed that one herder can easily take care of five thousand head," as reprinted in *Reno Weekly Gazette and Stockman*, March 12, 1896, 6.
34. Stabler, "Favorable Report on the Proposed Monitor Forest Reserve Nevada 1906," 9, Toiyabe National Forest.
35. Ibid., 10.
36. Ibid., 10.
37. Hatfield, "'We Were Not Tramp Sheepmen,'" 84.
38. Mark G. Woodruff to The Forester, Washington, D.C., November 20, 1907, Record Group 95, U.S. Forest Service, Files Office of the Chief, Grazing Plans, National Records and Archives Administration, College Park, Maryland.
39. Ibid.
40. Ibid.
41. Ibid.
42. Ibid.
43. Ibid.
44. Ibid.
45. Ibid.
46. Ibid.
47. Ibid.
48. Ibid.
49. Ibid.
50. Ibid.
51. Ibid.
52. Steen, *U.S. Forest Service*, 87, 163–64; Rowley, *U.S. Forest Service Grazing*, 60–66.
53. "Will Test His Authority," *Anaconda Standard*, September 11, 1908, 7; "Forest Reserve Suit Is Argued," *San Francisco Call*, September 11, 1908, 5; "Livestock Company Denied Injunction," *Los Angeles Herald*, October 9, 1908, 1; "Decision Against Stock Company and in Favor of Government," *Reno Evening Gazette*, October 9, 1908, 8; "Fights President in Forest Reserve Case," *San Francisco Call*,

October 9, 1908, 5; Steen, *U.S. Forest Service*, 88–89; U.S. Forest Service, Range Management Staff, *Court Cases related to Administration of the Range Resource on Lands Administered by the Forest Service*, 9–18; Rowley, *U.S. Forest Service Grazing*, 66–68.

54. Clawson, "Range Lands of Northeastern Nevada," 6; Lane, "Cultural Ecology of Sheep Nomadism," 98–111, 121–131; Ivan Sack, "History of Toiyabe National Forest," 1965, unpublished manuscript, Special Collections, Library University of Nevada, Reno, Nevada.
55. Frank V. Perry, "The Last Indian Uprising in the United States," *Nevada Historical Society Quarterly* 15, no. 4 (Winter 1972): 23–37; see also Effie M. Mack, *The Indian Massacre of 1911 at Little High Rock Canyon, Nevada* (Sparks, Nev.: Western Print and Publishing Company, 1968).
56. Wooton, *Public Domain of Nevada and Factors Affecting Its Use*, 33–40; Young and Sparks, *Cattle in the Cold Desert*, 95–98; Patterson et al., *Nevada's Northeast Frontier*, 310–11.

4

"Desirable Immigrants"

Socioeconomic Ambivalence and Basque Labor in Nevada's Sheep Industry, 1910–1939

In the summer of 1923, Angela Odriozola Larrucea (1903–2001), a young Basque immigrant woman, arrived in Elko in northeastern Nevada. Angela was born on September 15, 1903, in Gernika, Biscay. She grew up spending the days in the countryside, helping her mother with the housework, raising vegetables, and harvesting different types of fruits in her family's farmstead (or *baserri*) named "Cigarreta," located in the town's neighborhood of Saint Pedro. The young Angela went to America with the intention of soon returning home after she completed the task, assigned by her mother, of taking care of her aunt, who was living at that time on a ranch in Ryndon, Nevada.[1]

One year prior to her departure, with the help of her parents, Angela made all the necessary arrangements to meet American immigration rules. On June 7, 1922, Angela underwent a general medical examination by Rafael Echevarria, titular doctor of Gernika-Lumo. Echevarria attested that Angela was in a "good state of health." On the same day, Gabriel Toña, mayor of Gernika-Lumo, officially authorized the medical certificate. Toña also certified, according to the gendered social standards of the time, that Angela Odriozola dedicated herself solely to housework and had "good general aptitudes exercising home duties." The next day, before the district court judge of Gernika, her father, Pedro Odriozola, authorized Angela to go by herself to "California" and take care of her aunt. Even though Angela was leaving for Nevada, the affidavit read "California," because at that time in the Basque Country, California generally referred to the entire American West. Nearly a year later, on June 9, 1923, Angela got her passport from the Spanish authorities in the Civil Government of Biscay in Bilbao. On June 15, 1923, Lucius H. Johnson, American vice-consul in Bilbao, granted Angela a visa that would allow her to enter the United States under the Spanish quota, which at that time was set at 912 immigrants (however, even with a visa, immigration authorities in Ellis Island could still deny her entry if they determined to do so).[2]

On June 23, Angela departed for America from the port of Le Havre in France. A week later, she arrived in New York. When she went into the registry room in Ellis Island, the immigration authorities detained and confined her for a week for immigrant medical inspection in a barracks for newcomers. She possessed all the necessary documents to prove that she was healthy and had her father's legal consent that justified her purpose for travel. At that time, immigration inspectors feared that any woman immigrating alone without male protection would fall into prostitution and thereof could be denied entry. With her papers in order and just nineteen years old, Angela was allowed to enter the country.[3]

From New York, she took four days to travel to northeastern Nevada. On her arrival in Elko, Angela went directly to the Basque-owned Star Hotel, where she spent the night. The next morning Celso Madarieta, a Basque immigrant living in Elko, took her to the ranch where Angela's aunt lived. At first, Angela could not imagine living in this strange country. She cried for weeks. However, a month after her arrival, Angela met Gavino Aguirre, a Basque immigrant who at that time was working as a cowboy on a livestock operation near Ryndon. Four months later, on November 3, Angela and Gavino got married in Elko, where they raised a big family. Contrary to her first intentions and expectations, Angela soon adapted herself to her new life, loved it, and stayed forever in the United States.[4]

Angela Odriozola arrived in the United States at a complicated time for immigrants. The new immigration from eastern and southern Europe at this time was arriving, as historian John Higham has noted, during a high tide of xenophobia and nativism. "The new immigrants," as Higham wrote in *Strangers in the Land,* "had the very bad luck to arrive in America en masse at a time when nativism was already running at full tilt."[5] Opposition to the great numbers of new immigrants was also occurring on the western ranges at a time when the federal government had set in motion changes to regulate and even close some of the open rangelands. This placed Basque immigrants under a double threat as they struggled to overcome anti-immigrant attitudes and restrictions on grazing, especially large roaming sheep bands.

The present chapter analyzes the historical process of social and economic adjustments of the Basque immigrant community between the decades of 1910 and 1930. By 1910, the steady flow of immigrants from the Basque Country had guaranteed and strengthened the endurance of this community in Nevada. Nevertheless, as they became a well-accepted and respected immigrant group, Basques began facing greater legal constraints on their status as immigrants and on their engagement in itinerant sheep grazing. They would also face an onslaught of highly public attacks by Nevada's senator Key Pittman and the cattle industry, who viewed the Basque sheepherders and sheep industry in general as a threat to range resources. The Immigration Act of 1924 would sharply reduce immigration from southern and eastern Europe, including Basque immigration. For the sheep industry, this would mean a Basque labor shortage in the future.

The Growth of the Basque Immigrant Community and the Increasing Migration Network

In the 1910s, a settled Basque immigrant population in Nevada had served to recruit and attract the emigration of more of their countrymen. In 1910, the estimated population of Basque immigrants in Nevada, California, Idaho, and Wyoming was approximately 8,400.[6] As Basque immigration grew, a number of businesses appeared in the Basque Country to provide assistance to the immigrants, such as travel-related services, transitory hotels, and financial help. With the growing demand for sheepherders, labor agents appeared to capitalize on the Basque immigration by recruiting young boys and sending them overseas. Besides labor agents, travel agencies and banks also entered the picture. They often publicized their services by word-of-mouth as well as advertised in local newspapers and bulletins in the Basque Country. In a 1911 yearbook of the Basque Catholic Church of Bayonne, there were included some commercial advertisements, some of them related to immigration services, such as the following announcement of a travel agent named Paul Novion:

> PAUL NOVION
> 26, Quai Galuperie, 26

> IN BAYONNE
> Tickets for America in the so-called Messageries Maritimes steamships from Bordeaux to Buenos Aires, from Le Havre to New-York and California, from Saint-Nazaire to Vera-Cruz and Havana, from La Pallice to Chile. [writer's translation][7]

Furthermore, the Basque newspaper *Eskualduna* based in Bayonne, Labourd, became a referential service portal for those Basques willing to depart for America. During the 1910s, it was very common to find advertisements of immigration-related services in this newspaper, as in the following from the *Crédit Lyonnais* bank in Bayonne:

> CRÉDIT LYONNAIS
> In the Liberty Square of Bayonne
> Crédit Lyonnais Takes and Pays
> CHECKS FROM AMERICA
> and every accounting document.
> It also sells and buys every accounting document and title.
> Crédit Lyonnais provides clarifications on purchasing titles and the best advice to make all the arrangements properly.
> THERE IS ONE BASQUE MAN
> who is employee of Kredit Lyone and he is in charge of these issues
> HE SPEAKS BASQUE[8]

Also, in the same newspaper, it often advertised information concerning immigration procedures to the United States from France:

> TO AMERICA. Those who want to go to America address BARBIER, from Saint-Jean-Pied-de-Port. He is known as *Agent d'émigration* of the Government. He needs to take the whole day to undertake these duties in Bordeaux, and he assists passengers as they embark on the ship.[9]

On the other side of the Atlantic, in the Far West, Basque-owned boardinghouses served as employment centers, primarily for sheepherding. In the 1910s, each Nevada town had at least one Basque boardinghouse. These establishments appeared and spread all around Nevada, which turned this state into the leading Basque immigrant center in the West. These Basque-owned establishments were the first separate facilities to provide newcomers with all kind of assistance: cheap beds, hot food, laundry, storage services, health care, or mailbox facilities. In addition, it was a place where they could speak in their native language and not feel intimidated by others when using their language. Typically, the hotels were located near the railroad station to make it easier for newcomers to reach these establishments upon their arrival. In 1910, for instance, Basque immigrant Pete Jauregui built the Star Hotel, situated across the railway from the Western Pacific depot in Elko. Essentially, these houses were immigrants' first refuges in the Anglo society. They functioned as nuclei where Basque immigrants could start and set up their new lives in the West.[10]

By the 1910s, more commonly, Basque hotelkeepers advertised their establishments in the local newspapers of Nevada, announcing them as cozy places to enjoy good food and rest in a house-keeping room, as well as employment service offices. In the early 1910s, an advertisement of the Overland Hotel in Elko owned by Domingo Sabala and Eutalie Onaindia was repeatedly run unchanged in the *Elko Daily Free Press*. The advertisement read as follows:

> D. D. Sabala E. ONAINDIA
> Overland Hotel
> Fine Wines, Liquors, and Cigars First-Class Restaurant
> Private Wine Rooms Rooms Electrically Heated
> The Only 3-Story Fire-Proof Hotel in
> in Eastern Nevada. A splendid view
> of Elko from our 3-story building.
> Free baths to guests.
> Headquarters and Information Bureau for sheepmen and
> sheepherders.
> ELKO, NEVADA[11]

Significantly, Basque boardinghouse owners, like Sabala and Onaindia in Elko, advertised their businesses in the local newspapers to a broad and diverse clientele, even outside the Basque community borders.

Boarding houses became important contact zones between the Basque immigrants and the wider society. As Basques became more integrated and familiarized with the American environment, non-Basque people were welcomed and regularly began frequenting these establishments. Visits by non-Basque guests in these hotels promoted contacts by the Basque immigrant community with the larger society. For multiple reasons, these boardinghouses were the most important refuge institutions for Basque immigrants and ultimately provided a primary means of Basque assimilation into American culture and society.[12]

Because of occupational gender roles, Basque immigration was predominantly male as the men took jobs in stock operations. At the turn of the century, however, Basque women began to follow their male counterparts and emigrated to the West as well. Typically, Basque immigrant women went into housekeeping and other jobs helping sheepherders from different locales: managing boardinghouses, cooking for the ranch hands, or helping their family members in the new country. This was the case of Jesusa Guridi (1890–1982), who in the summer of 1916 emigrated from the small village of Ea, in Biscay, to McDermitt, on the Nevada–Oregon border, to join her father and three brothers. There, she began working as a cook on the Sullivan Ranch. The opening of new Basque boardinghouses led to an important increase in the demand for domestic work. As these establishments stimulated a demand for help, Basque women were ready to take up their roles in the Basque immigrant community.[13]

Working in the boardinghouses not only gave Basque women a new social status, which they hardly could reach in the Old Country as housekeepers in their families' houses, but also gave them the opportunity to participate actively in the public life of the host society. In the Basque Country, women were traditionally confined to matters of home and family. Peasant women typically were constrained to labor with other family members in agriculture in the family's farmstead, as well as caring for the house. On the other hand, in the United States, although they were confined to traditional women's occupations, they found paid work outside the home. For example, in 1927, Catherine Goyhenetche Etcheverry arrived in Eureka, Nevada, and was first employed at the boardinghouse of the Uriarte family as a cook. Some months later, she started working in the Eureka Hotel, owned by the Laborde family, for much better pay. These types of occupational opportunities did not, however, necessarily mean that Basque immigrants broke with old societal standards. Ordinary individuals entering into the unsteady society of the American West, as historian Anne M. Butler has pointed out, often "felt the need to reaffirm old values, cherish them more dearly, and impose them more stringently." This was true of the Basque women in Nevada, where hard work and perseverance were needed in order to make their new living. In general, Basque women played a prominent role in preserving their culture, diffusing it, and constructing associations or relations with the members of the wider society through, for example, their participation in the Catholic Church, a significant religious and cultural institution in Basque life in Nevada.[14]

In the New World, Basque immigrants used the Catholic Church as another refuge from which to draw strength as they adjusted to their new environment and work. The Church offered traditions from the Old Country and at the same time assisted in Basque assimilation and integration. Since the early days of statehood, religion was an important part of life in several rural and mining towns in Nevada, something that would intersect with the Basques' own Catholic traditions, which they brought with them from Europe. Since the 1860s, the ethnic and religious mixture of Nevada, similar to that which occurred in other western states, was reflected by a cultural pluralism in the sectarian identification. Catholics, Protestants, Mormons, and Jews were widely present by the late nineteenth century, evidenced by the construction of churches in the main towns all over the state. During the Nevada's first mining boom in Virginia City, a silver mining boomtown, the Irish supplied much of the labor in the emerging mining industry. Irish immigrants brought with them their culture and social institutions, including their Catholic faith.[15] The Irish were pivotal in developing the infrastructures and social networks for the development of Catholicism in Nevada. Irish settlers were followed by other Catholic immigrant communities: Italians, Swiss, Slavs, Portuguese, and Basques. All of them formed the so-called "church of the immigrants" in Nevada.[16]

At the turn of the century, Catholics—either priests or women's religious orders—worked to make their church a religious force in the American West. The Catholic Church of Nevada was a protective institution for various immigrant communities. The Church also served to integrate newcomers to their new country. In the early twentieth century, furthermore, some Catholic Churches in Nevada provided services expressly for the local Basque immigrant community. There were some Basque priests settled in the United States who made stays in the Nevada towns and even visited the ranches, preaching to the Basque immigrants. In 1910, for example, Basque priest Gratian Ardans who belonged to the Order of St. Benedict at Sacred Heart and founding Pastor of Our Lady of Lourdes Catholic Church in East Los Angeles,

California, conducted services in English, French, and Basque in Reno. Gratians and other Basque pastors continued visiting the Basque community of Nevada throughout the early twentieth century.[17]

Being a Catholic in an Anglo-Protestant world, however, did have certain drawbacks, and Basques were subject to ethnic and religious-based prejudice in many rural frontier situations. Nonetheless, the Catholic Church was also able to serve as a social shelter in defense against this prejudice. As Basque immigrants reemphasized their Catholicism, they also found it to be a vehicle for social mobility. Basques were active members and contributors to the communal and social welfare work of the Church. The Altubes, who at the time were already a wealthy family with one of the largest stock operations based in Nevada, made large contributions to the Spanish Church in their home base in San Francisco.[18]

Their Catholic affiliation gave the Basque immigrant community a public voice and an opportunity to strengthen their ties with the overall society. For example, when Key Pittman was campaigning in Nevada for the U.S. Senate on an anti-Basque, anti-sheepherding platform, Father Thomas M. Tubman, Reverend of St. Thomas Aquinas Cathedral of Reno, wrote a letter to the editor of the *Reno Evening Gazette* that was published on March 31, 1911, in support of the newspaper's response to Senator Pittman's outbursts against the Basques and the sheep industry. In this letter, Father Tubman expressed his admiration for the Basque immigrant community and their support of the Catholic Church. The letter reflects the Church's embrace of the Basque community. He wrote:

> The Basques who are now here have made good. They have shown how wealth can be wrung from our uncultivated plains and rugged mountains. With indomitable energy they have gone to work in raising great sheepfolds, that have brought wealth to themselves and the state.
>
> They are admirably adapted to pastoral and agricultural pursuits, strong of limb, healthy in body, temperate in habit, religious and Godfearing, there can be no failures among them.[19]

Tubman not only praised the contribution of the Basque immigrant community in Nevada, but also deplored the discrimination they suffered because of their work on the rangelands as sheepherders. In his praises and claims in favor of the Basques, Tubman's letter shows how by the 1910s, the Basque-American community was not only settled in the sheep camps of the Great Basin, but also in towns like Reno.[20]

By the early 1910s, although still faced with discrimination and stigmatization on the ranges, it can be said that Basque immigrants were settling and integrating into the broader American society of Nevada and other western states. On March 29, 1911, the *Reno Evening Gazette* published an article with the headline: "Desirable Immigrants." It noted approvingly that the Basque population in Nevada was "increasing year by year until they now form no inconsiderable proportion of the inhabitants of the state." Curiously, the Reno newspaper explained that the origins of the Basque "race" remained unknown, by stating that "neither their language, customs, skull formation nor their history nor traditions throw any light upon the subject." It further added: "All that is known of them is that they are an unconquered people, rendering tribute to no potentate, acknowledging the authority of no constituted power outside of their own community." From a typical environmental determinist position, moreover, the Reno *Gazette* attributed the Basques' labor in the western open-range sheep grazing to the immigrants' rural lifestyle and pastoral occupation in their home country: "Owing to their mountain life and their pastoral occupations, the Basques are a liberty loving people. Hence it is that they are so strongly attracted to America and particularly to the plains and mountains of the west."[21]

In terms of a classical American self-made ideology, the article praised Basque immigrants and accorded them a respected and honored position within the broader society because they had built up wealth through their labor in the sheep industry. It contended that nearly all the Basque immigrants became prosperous through their labor in sheep and ownership of land, just as their former employers did before: "They are becoming the heaviest owners of sheep in the State of Nevada, their ranges extending far and wide. Industrious and economical to a degree, there are few Basques in Nevada who fail to attain a competency within five years after coming to the state."[22]

Then, the *Gazette* argued that Basque immigrants were eagerly adapting themselves and assimilated quickly into mainstream American culture: "It is always their ambition to become naturalized as quickly as possible and, being a people opposed to race suicide, they marry early in life and rear large families. . . . The second generation of Basques . . . are most thoroughly Americanized and become perfectly assimilable." The article further praised the Basque immigrant community of Nevada as respectable and good citizens. In Nevada's "large colony of Basques," many of these immigrants had "amassed considerable fortunes. They are all law abiding, courteous, intensely religious and—what is most to the point—their aim is not to acquire sufficient money to take them back to the Pyrenees to reside for the balance of their lives, but to build a roof tree and rear their families in the land of their adoption."[23]

Despite this notably optimistic view of Basque immigrants, the realities of assimilation would have certainly been difficult and complex. In 1918, the Basque priest, linguist, and writer, Resurrección María de Azkue (1864–1951), published his novel in Basque language *Ardi Galdua* or "The Lost Sheep." This book tells a story through an exchange of letters between a Basque immigrant in California and his relatives in the Basque Country reflecting the social, economic, and religious issues of migration.[24] "*The Lost Sheep*" metaphor is especially noteworthy because it captures the alienation and displacement that many of the Basque immigrants would have felt upon their arrival in the Far West. Indeed, Basque immigrants who reached the United States would have needed to continually find ways to adjust and learn how to live and relate with others in the complex social dynamics of their new surroundings.

From "Dark-Faced Strangers" to White Subjects: Basque immigrants and Their New Racial Identity during the Restriction Period

The increasing presence of Basque sheepherders in the public ranges had drawn derision amidst a cowboy culture in the American West. By the early twentieth century, nomadic sheepherders and Basque immigrants were synonymous. Basque sheepherders with their huge bands of sheep coming onto public grazing lands were perceived as a serious economic and environmental threat by local ranch owners, small town business communities, and the emerging conservation movement. Anglo-Americans frequently denigrated and mistreated Basque sheepherders for their alleged disruptive and destructive utilization of the range resources.[25] On May 4, 1898, the Carson City *Morning Appeal* had alerted its readers that over forty Basque immigrants divided into two labor groups were bound for Nevada to work as sheepherders. The *Appeal* considered the upcoming arrival of these Basque immigrants as "Nevada's Danger."[26] In 1899, the influential *California Cultivator and Livestock and Dairy Journal* had contended: "The sheepmen of the Sierra Nevada are for the most part a lot of irresponsible Basques, who own no other property than their sheep; they pay no taxes, evading them by moving."[27]

After 1905, when the U.S. Forest Service began establishing a range control program in newly created national forests, the sheep industry's transhuman or migratory use of grazing resources on the public domain and the foreign character of its Basque workers attracted more attention from policymakers and public opinion.[28] American literature made Basques increasingly popular as outsider nomadic sheepherders in stories about grazing conflicts and pasturage problems. In *The Flock* (1906), Mary Austin, while describing sheepherders' experiences in California's Sierra Nevada, wrote that "The actual management of a flock on the range is never a 'white man's job," but French, Portuguese, Mexican, and Basque immigrants.[29] These collectivities that were considered nonwhite (including the Basques) were perceived as marginal people who were unable to find better employment and accepted this low-paying work readily.

The negative collective image of Basques as "tramp" sheepherders persisted for many years and was perpetuated by some cattle ranchers as a pretext to put the sheep industry out of business.[30] As historian Richard W. Etulain has explained, the widely held image of the Basque as a wandering landless sheepherder gave the entire Basque immigrant community a somewhat negative reputation.[31] Pejoratively they could be called "Greasers," or "Garlic Snappers."[32] Some years later, for example, Timotea Yraguen Echanis, a second-generation Basque-American woman, explained how when she first started school in McDermitt in the early 1920s, she faced humiliation and discrimination for not knowing any English but only Basque, and that she and her Basque classmates were oftentimes called "Black Bascos."[33]

Despite being praised in some quarters and in the press, Basques were still often seen as marginal and strange people, as well as a nonwhite race.[34] On April 7, 1907, an article appeared in the *Washington Post* describing the "arduous" life of the Basque sheepherder out on the range. On the matter of race, it depicted Basque immigrant sheepherders as being a nonwhite race. The article described Basque sheepherders as follows:

> There must be a hundred of these dark-faced strangers from Spain who lead the sheep through the Sierra range from Shasta down to Kings River. The valley folk call them gypsies in their ignorance.

> They are considered shiftless, roving fellows. Because they do not care to talk or to mingle with "white men" they are looked upon with suspicion. Yet everybody concedes that they are good shepherds. No other man would work all summer alone for so little pay.[35]

The article continued, explaining the seasonal migration of the transhumant sheepherders between lowland winter and upland summer pastures. It also pointed out that the loneliness in the rangelands invoked alienation and sometimes despair in these immigrants.[36]

Later on, in the context of World War I, a period of intense nativism and xenophobia in the United States, Basque immigrants and their cultural traditions led to a further demonization and "foreignization" of the sheep industry.[37] While Basque sheepherders continued to be seen as intruders by a large segment of an old established ranching community, on the other hand, Basque immigrants were increasingly identified as important contributors to some local communities. Although both negative and positive conceptions about Basque immigrants continued to coexist, those positive images would eventually become dominant throughout different western communities.

Key Pittman's Campaign against the Sheep Industry of Nevada, 1910–1920

By the 1910s, although an increasing number of Basque immigrants owned vast tracts of land, the idea of the ubiquitous Basque transhumant and landless sheepherder in the American West continued to resonate in the American public imagination. Such a caricaturization served to justify and protect the economic and political interests of the establishment at the expense of the Basque sheepherders. Nourished by increasingly xenophobic responses to immigration, some voices were raised against the Basque immigrant community in Nevada.

Some political personalities attempted to put forth solutions and even opposition to the presence of itinerant Basque sheepherders on the Nevada ranges. For example, Nevada's senator Key Pittman worked tirelessly on the range issues, particularly on the persistent problems affecting the cattle business in Nevada. His early political career was marked by continuous defense of the cattle community and a hostility toward the sheep industry. Pittman's efforts to extend the boundaries of national forests, eliminate the protective tariff on wool, and support of the

1916 Ranch Homestead Act all aimed to undermine the sheep business in Nevada. Furthermore, Pittman's anti-immigrant sentiments fitted well with his fight against the sheep industry that was synonymous with the Basque immigrant community.

Pittman against the Basque Community

In 1902, Key Pittman had served as a member of the Central Committee of the Silver Party of Nevada. In 1908, Pittman resigned from the declining Silver Party to join the Democratic Party. In the context of the national progressive reforms, Pittman launched an unsuccessful campaign for the Senate. While endorsing the Democratic Party's leading reformers, Pittman campaigned in favor of a reduction in the tariff rates, advocated to enlarge the scope of the Sherman Anti-Trust Act, defended a powerful national bank system, and endorsed federal irrigation projects. At the same time, Pittman raised his voice in favor of the private landowners and livestock ranchers, particularly cattle owners, often in a way that would be at the expense of Basque immigrants.[38]

In the early twentieth century, the continuous expansion of open-range itinerant sheepherding posed a threat to the economic interests of some cattle and sheep companies using public-domain lands in Nevada. By then, nomadic bands of sheep from Utah and Idaho wintered in the eastern part of Nevada in White Pine County. Typically, these nomadic bands of sheep were owned by sheep companies who employed cheap immigrant labor, especially the Basques. During the 1910s, Pittman launched a campaign against the open-range sheep industry in Nevada. Basque workers became the focus of Pittman's attacks. He blamed Basque sheepherders for depleting public ranges. Nevertheless, the Basque community would wither the attack and even gain broad support.

Key Pittman's hostility toward Basques did not go unnoticed. His blustery statements captured the attention of the media and the public in Nevada and other western states where the Basque population concentrated. On October 7, 1910, the *Reno Evening Gazette* questioned Key Pittman's attacks against the sheep industry and the Basque workers. The *Gazette* complained about Pittman's allegations as follows: "Some good Democrat ought to tell Mr. Pittman that he is making a joke of himself when he campaigns along those lines." The Reno newspaper considered the senator's statements unfair, and because the sheep industry represented an important sector in the state's economy, the *Gazette* saw Pittman's position as being incomprehensible: "Upon the ranges of Nevada today there are running about one million sheep. The average price realized from a fleece is over one dollar. The income from sheep sold as mutton is over two million dollars. Over three thousand sheep men are employed directly and indirectly in this great industry."[39]

Furthermore, it declared that Nevadans were intelligent enough not to be swayed by these kinds of arguments and "to distinguish between fit and unfit—and they know that any man who would destroy the sheep industry of this state simply because he believes or pretends to believe that the only men who benefit from it are twenty sheep owners and two hundred Basque sheepherders is unfit to look after Nevada's interests in the United States Senate."[40]

Despite these newspaper pieces, in January 1913, Key Pittman was elected a senator of Nevada. In this position, he continued to advocate for his cattle-grazing constituency. On June 4, 1913, Pittman testified before the Senate Lobby Investigation Committee in Washington, D.C., which was carrying out a general investigation to determine the proposed tariff bill. Representing the Nevada Democrats, Pittman defended the reduction of the tariff rate on wool arguing that the sheep industry of Nevada was one of the less important sectors in the state's economy. Besides saying that the sheep industry amounted to little, Pittman remarked that a large percentage of the total sheep population—over a million sheep—were owned by out-of-state owners from California, Idaho, and Utah. Pittman continued: "I also know of my own personal knowledge that nearly all of the laborers employed in that occupation are Basque sheep herders, who can hardly speak the English language, and get about $35 a month."[41]

Immediately, James A. Reed, senator from Missouri and member of the Committee, got interested in the question of the Basque immigrant labor:

> SENATOR REED: What do you mean by Basque sheep herders?
> SENATOR PITTMAN: The Basques come from the Pyrenees mountains in Spain. They get that class of labor because they seem to be adapted to sheep herding, and they are lacking in intelligence, independence, or anything else. They are just about as near a slave as anybody could be under our present existing conditions, and I do know of my own personal knowledge that the sentiment of our State is not in favor of a tariff on wool or sugar.[42]

Pittman denigrated the Basque immigrant labor as inferior and despicable. He blamed Basques for being an economic threat to Nevada because they took these agricultural jobs and depressed local wages. A brief time later, Reed asked Pittman more about the Basques of Nevada:

> SENATOR REED: You spoke a moment ago of the Basque sheep herders. Are they ordinarily naturalized citizens, or otherwise?
> SENATOR PITTMAN: No, they are not. As a general thing they never associate with the other people in the State; they live among themselves; they can only speak a few words of the English language; they live in the lowest possible way for a human being to live; and they are nothing but sheep herders.
> SENATOR REED: You spoke of the wages, $35 a month. Is that an ordinary wage in your State, or is it less than the wage that other people receive?
> SENATOR PITTMAN: It is much less.[43]

According to Pittman, the Basque immigrant community of Nevada was very clannish and unassimilable. He sneered at Basque Americans because he erroneously considered them as mere sojourners. To Pittman, in other words, the Basques were undesirable immigrants many of whom would never become citizens of the United States.[44]

Pittman's declarations infuriated not only the Basque immigrant population living in the West, but also other non-Basques who stood up publicly for the Basque residents. Public reactions to these statements came first from the non-Basques. On July 24, Nevada attorney J.A. Langwith from Winnemucca wrote a letter to the *Silver State* criticizing Pittman's accusations and paying a special tribute to the Basque community. He said that Pittman's declarations had caused a loss of confidence among businessmen in other economic sectors of Nevada: "...Pittman's attack on the sheep industry and the Basque sheepowners and herders is doing much injury to our state. Business men and men with money are figuring that our leading industries, such as cattle, sheep, lead, mining, and possibly silver and copper, will get some very hard knocks should cattle, wool and lead be eventually placed on the free list." Langwith also questioned the senator's charges against the Basque immigrants, considering them as untrue and unfair, and he stuck up for the Basque community of Nevada: "There are many Basques and descendants of Basques engaged in ranching, cattle and mercantile business as well as the sheep business. The Basque is usually a man of more than ordinary intelligence, with keen business ability, and is usually prosperous. They are, as a rule, law abiding, sober, and make good citizens." He explained how Basque immigrants were not only mere sheep workers, but were also engaged in every level of the livestock industry, whether cattle or sheep. Langwith enjoyed good relations with the state's Basque immigrant community, and he could not understand Pittman's attitude toward them: "I feel it my duty to thus poorly express what I know about the men and women whose fair name, honor, pride and integrity Senator Pittman has so maligned and deeply wounded." The dignity of the Basque sheepherders and the overall immigrant community became a constant issue in letters and articles appearing in numerous newspaper during this time.[45]

During the debate of the tariff bill, on August 21, 1913, in the first session of the sixty-third Congress, Pittman gave a long speech on the floor of the Senate, declaring himself "heartily" in favor of the bill H.R. 3321 to reduce tariff rates on wool and that he was "firmly convinced that it is for the best interests of the people of my State." Pittman responded that lowering tariff rates for wool did not mean fiercer competition for domestic sheep companies:

> The removal of the tariff will certainly permit foreign wool to come into the country, but the question is: How much will come in, and at what price can the foreigner afford to sell it in the United States?
>
> In the first place, the foreigner can not afford to sell his wool in the United States for less than he can obtain for it in the markets of the world; and in the second place, he can not afford to sell it for less than it cost him to produce.[46]

To Pittman, the problem depended on the balance between the supply and demand: "With the demand steadily increasing and the supply rapidly decreasing the competition must be between the buyers instead of the sellers. In such event, the price would be so high that the cost of production in foreign countries would not concern the American woolgrowers."[47]

After arguing the benefits of tariff liberalization, Pittman contended that the sheep industry in Nevada was insignificant compared with other western states, and that cattle ranching and the mining industry were more important for the state's wealth. Pittman advocated the protection of settlers and home builders against sheep grazing in the public range in Nevada. According to Pittman, sheep grazing was detrimental for the agricultural development and farm settlements in Nevada. In strongly nationalist terms, he described the following disadvantages of the sheep industry to Nevada:

> The American farmer comes naturally to the raising of cattle and horses, and he has no superior on earth, while the sheep industry, with its cheap labor, seems to require the most ignorant, the most unprogressive, and the lowest type of foreign labor. It is the custom of the farmer in our valleys to range his cattle and horses on the adjacent mountain side while raising and harvesting his hay, and then to drive them within his inclosure, fatten them upon the grass and the hay, and drive them to market. Since the sheep industry has monopolized the range of the State of Nevada the farmer finds it difficult to pursue this system of raising cattle and horses. Down each side of the valleys, along the mountain ranges adjacent to these farms, come thousands upon thousands of sheep, driven by Basque sheep herders and collie dogs, uprooting the vegetation, breaking down fences, destroying roads, obliterating ranges, defiling the watercourses, and driving the cattle and horses of the farmer off of their natural ranges.[48]

While demeaning the sheep industry, Pittman stood up for homesteading and the cattle industry, saying that there was "ample room to increase the number of farms and to increase the number of the farmers' cattle and horses if the sheep are not permitted to monopolize the public domain, the springs, the wells, and the watercourses of the State."[49]

At the end of his speech, Pittman said that "practically all of our sheep herders and nearly all of the laborers engaged in the sheep industry are Basque herders . . . who speak very little of the English language and rarely ever declare their intention to become citizens of the United States." In strong sympathy with a pro-labor progressive ideology in Nevada, Pittman contrasted itinerant sheep laborers with the stalwart unionized immigrants who worked in the hard-rock mining industry of his state:

> As to the employment of foreigners in other branches of labor in my State I wish to say that the other foreigners who are engaged in labor in the State are engaged principally in mining; not coal mining, for we have none, but hard-rock mining. There are no higher class laborers than miners. All of them are union

> miners. All of them stand for union wages. They are all capable, intelligent workers, and every one of them declares his intention to become a citizen of the United States just the minute the opportunity is offered to him. There is no comparison whatever between the ordinary foreigner and the Basque sheep herder from the Pyrenees Mountains.[50]

On October 3, 1913, Congress finally passed the Underwood Tariff Act which placed raw wool on the free list. Western woolgrowers were angered with the passage. They were particularly furious with Key Pittman's declarations about the sheep industry of Nevada. In its November issue, the editorial of the *National Wool Grower* (the organ of the National Wool Growers Association [NWGA]) refuted each of Key Pittman's charges against the sheep industry of Nevada. The National Wool Growers Association found Pittman offensive and contended that everything he had said about the sheep business in Nevada rested on vague generalizations. The *National Wool Grower* complained that Pittman was making false accusations against Nevada's sheep economy, claiming that this industry represented more than 25 percent of the total value of the entire livestock economy of the state. Also, the editorial of the *National Wool Grower* contended that 1,154,000 sheep were owned by ranchers from Nevada. The editor said that Pittman's accusations of the Basques were "untrue," arguing that:

> Less than one-half of the labor in the sheep industry of Nevada are Basques. But suppose there are some Basques in Nevada who happen to be in the sheep business. Does it necessarily follow in the senator's opinion that because a man is foreign born he is necessarily an undesirable citizen? The census shows that 25 per cent of Nevada's entire population is foreign born. We happen to know, however, many Basques in the sheep business who are honorable citizens and several Basques in Nevada have contributed more to the upbuilding of that State than Senator Pittman ever did or would if he lived there one hundred years.[51]

Moreover, the editorial of the *National Wool Grower* said that Pittman's statements on the wages were baseless and "downright falsehood." The *National Wool Grower* reprinted some statistics of the Tariff Board in order

Table 4.1. Wages of sheep labor

Class of labor	Pay in dollars per month
Superintendent	111.5
Camp tender	64.93
Herder	59.42
Extra labor	63.02

Source: National Wool Grower 3, no. 11 (November 1913): 23.

to contradict Pittman's statement on low wages, revealing that the pay of the sheepherders was about $60 (table 4.1).[52] In addition, the *National Wool Grower* said that Pittman's "talk about sheep herders being slaves is unworthy of the consideration of intelligent citizens." Further, the NWGA considered Pittman an outsider or carpetbag senator who was not interested in the prosperity of Nevada. It noted that Pittman had investments himself in Alaska, where indeed he had moved from, and that he was doing much more for the far northern arctic state than he was doing for Nevada.[53]

Understandably, Pittman's statement was not well received by the Basque community. In October 1914, with the Great War as a backdrop, the *Reno Evening Gazette* published a letter by a writer who identified himself as a Basque sheepherder and who argued strongly against Pittman's position. "Nine-tenths of them [Basques] own their land and their homes, dependence being considered the lowest condition of social life, and none being ever excused from work, that has a mind to think and a hand to wield . . ."[54]

Similar expressions of indignation against the attacks on the Basque community and the detrimental effects of the low tariff bills on wool appeared in other western states. From Boise, on October 27, 1916, Sol Silen, one of the most influential spokesman for the Basque immigrant community, wrote a letter to the editor of the *Reno Evening Gazette*, in which he bitterly expressed his displeasure about Pittman's attacks on the Basque immigrant community. Silen complimented Basque immigrants: "Wherever the Basques choose to establish their home, that community is the better. They are an intelligent, energetic and progressive people: they hold valuable landed estates, erect substantial business structures and homes that are a credit to their respective communities." Silen's argument praised the Basque "race," its achievements, and historical heritage in America.[55]

Advocating for the Enlargement of National Forest Lands in Nevada

Pittman blamed Basque sheepherders, whom he sometimes called "tramp herders," for creating problems on the public grazing lands of Nevada. While the U.S. Forest Service had taken control of some parts of the grazing situation on the public lands, especially in the high-mountain meadows, a vast open range still remained outside Forest Service management. Range conflicts persisted on these lands. After his election to the senate seat in 1913, cattle ranchers appealed to Pittman to find ways to curtail itinerant sheep operations. Pittman, as a Democrat who supported placing wool on the free list, and as a supporter of the Forest Service, advocated the extension of its regulations over larger parts of the public domain to exclude an increasing number of newcomers with their herds coming onto public grazing lands. Pittman not only defended the limited number of national forests in Nevada, but wished to expand them so that they might regulate and exclude sheep from larger parts of the public domain.[56]

With the establishment of the Forest Service in Nevada as discussed in the previous chapter, studies had been begun on forage resources in relation to livestock grazing in order to determine an optimum economic use of the ranges. The Forest Service evaluated and made decisions on range management based on the "carrying capacity" of the ranges as determined by the availability and conditions of range forage. The Forest Service's implementation of grazing fees associated with the issuing of permits on the number of stock grazed generally favored cattle permits, but sheep operations were not excluded, although there were complaints to the contrary. During and after World War I, state and national representatives became increasingly irritated and frustrated about the uses of the still-open rangelands. The Stock Raising Homestead Act of 1916 tried to promote homesteading on the lands with parcels of 640 acres, but only resulted in further monopolization of the lands by large outfits. The amount of land was still too small for successful small homestead ranching.

In the late 1910s, especially after the United States had entered World War I in April 1917, the state of Nevada witnessed a remarkable expansion of the livestock industry, largely based on the wartime demand for wool, mutton, and beef. High prices for livestock meant prosperity. The Forest Service, for its part, accommodated its regulatory policy to the wartime economic demands and relaxed its protective limits on numbers of stock. High stock prices and almost open admission to national forest range resources quieted range conflicts at least for a while in rural

Nevada. During that period, however, an increasing number of newcomers arrived with new sheep bands to share the resources. At the same time, established ranchers, be they sheep people or cattle people, raised concerns about noncitizen immigrants possessing grazing permits in the national forests.[57]

After the end of the war, there followed a severe depression in the livestock industry and prices and production dramatically declined. Consequently, many ranchers were forced into bankruptcy. From 1919 to 1922, the postwar crisis was followed by a slow recovery because of a credit crunch which had a negative effect on capital investment in the livestock industry, characterized by a reduction of competition and increase in risk.

During this time, disorder in the sheep industry persisted. In 1923, the Nevada Legislature passed a livestock branding act through a system of registration of stock brands under the supervision of the stock inspector. Although the new state branding law served to discourage sheep itinerancy, conflict continued in the public grazing lands beyond the Forest Service's national forests in Nevada. In Grass Valley of central Nevada, on July 9, 1924, Thomas E. Brackney, a prominent rancher of Austin, killed a Basque sheepherder employed by the Sabal Estate and Sheep Company for allegedly trampling his ranch property with his flock.[58]

In the early 1920s, responding to his cattle-grazing constituency, Pittman moved to exclude immigrant sheep operators from using the public grazing lands. Some of his constituents had written Pittman about expanding Forest Service regulations to the public-domain grazing lands outside the national forests in order to discourage "tramp" sheepherding. In February 1924, Pittman wrote Chief Forester William B. Greeley, advocating for the enlargement of the national forest lands in Nevada in an attempt to have the Forest Service extend its grazing regulations over the public-domain range with the intent of further restricting itinerant sheepherding. The Forest Service, however, denied Pittman's request to expand national forest jurisdiction in Nevada on the grounds that "the timber within the proposed reserve was not sufficient to warrant conservation, and that such timber did not serve sufficiently to protect a watershed of useable flowing surface water."[59]

Pittman attempted to persuade Greeley to find a legally acceptable way to enlarge the national forest lands in the state of Nevada. Pittman justified the need to expand the existing regulatory system on economic efficiency grounds: "The use of forest reserves in Nevada for range control has proven of great benefit to the stock growing industry. Range control is essential in Nevada if the range is to be preserved. There is not range control legislation for the unreserved public domain and therefore it becomes important for stock growers to utilize wherever it is beneficial, range control through forest reservations."[60]

Despite Pittman's perseverance, the Forest Service eventually denied the petition of the senator from Nevada, suggesting that Nevada's lands were unworthy for further protection. However, the Forest Service empathized with Nevada's stock owners' struggles: "During the past ten years, however, transient sheep owners, excluded from other ranges by private acquisition, or otherwise, have come into this region in increasing numbers and have not only deprived the local stock growers of the forage resources upon which they are dependent but, through excessive grazing and improper use of the range, have largely destroyed the grazing value of the lands."[61]

Although the conservation movement and private companies shared a perspective on safeguarding and preserving economic stability, this time the Forest Service considered unjustifiable the expansion of its lands—or, more to the point, it did not believe it had the lawful authority to do so. Declaring it beyond their powers, the Forest Service and the secretary of agriculture judged necessary that Congress should enact legislation for constructively managing the use of the public-domain lands for grazing purposes.[62]

Nevada cattle owners, nevertheless, became increasingly frustrated over government inaction concerning the management of the public-domain lands in Nevada. In April 1924, the secretary-treasurer of the Quinn Canyon Livestock Association wrote to Pittman complaining about the overgrazing and destruction caused by the sheepherders in the public grazing lands beyond the national forests. He wanted some form of government control or regulation to stop the so-called tramp or transient sheep graziers because, in his own words, they had "absolutely no regard for the welfare of our ranges, his only interest being to avail himself of the maximum results and benefits of our ranges so long as they last." Much to the disappointment of cattle ranchers, Pittman could not obtain legislation to permit the expansion of national forests, and the Forest Service and the Department of Agriculture declined to cooperate with him in the effort.[63]

Post-World War I Immigration Restrictions and Labor Shortage in the Sheep Industry

In the decade following World War I, Congress reacted to the strong nativist attitudes that had gained momentum in the United States. Between 1917 and 1924, Congress adopted measures to restrict immigration from Europe, disrupting Basque immigration to the United States. In February 1917, before the United States entered World War I, Congress passed an immigration act that established more rigid criteria for admission and strengthened mechanisms for excluding some nationalities. The 1917 Act implemented the following provisions: it established a literacy requirement, authorized deportation, enlarged the list of deportable aliens, and increased the categories of Asian exclusion. Although this act did not hinder significantly the increasing recruitment of the Basque immigrant labor, it made it more difficult for those Basques already in the country to obtain naturalization.[64]

The 1917 immigration act did provide a legal avenue to permit the entrance of immigrant labor. It kept the doors open to "skilled" workers, among others, "if labor of like kind unemployed can not be found in this country." The act mandated the secretary of labor to determine "the necessity of importing such skilled labor in any particular instance." In the near future, when a Basque labor shortage became increasingly acute (particularly in the late 1930s as a consequence of the increasing immigration restrictions from the Immigration Act of 1924), this provision would serve as a legal mechanism to allow the recruitment of more Basque immigrants, causing those in support of Basque immigration to seek to demonstrate that Basques were the only "skilled" workers for the open-range sheep industry of the West.[65]

Although the labor shortage would worsen in the coming decades, in the war context of 1917, a sheepherder immigrant labor shortage was already becoming apparent. From Fallon, Nevada, on September 28, William A. Keddie, a prominent Nevada rancher who at that time ran the Williams family outfit, wrote Pittman indicating that there was a critical labor shortage in Nevada's sheep industry: "This year we have experienced considerable difficulty in securing sheep-herders and the men for ranch work. This shortage of help has without a doubt lessened the amount of our products." Keddie noted that this labor shortage partly was because of some "BASQUE [*sic*] herders we now have and who are becoming entirely too independent to be useful, to say nothing about the growing scarcity of getting them at all." As Basques quit sheepherding jobs and found other better paid employment and became U.S. citizens, Fallon-area ranchers looked for other immigrant groups to recruit dependable labor, particularly Romanian and Armenian immigrant workers. Keddie asked Pittman to find ways in Congress to open the doors for these immigrants, "securing 50 to 200 of these folks, preferably those from a sheep raising country of theirs." Keddie further noted: "Our idea was to get a number of these people, those without families preferred, use them on our ranches until we had accustomed them to our ways etc. and then break them in for work on the ranch and with the sheep."[66]

Keddie's statement exemplifies a determination on the part of ranchers to obtain immigrant sheepherders (young and single from Europe) and make them permanent sheepherders. Pittman also took up the matter with the Department of Labor and the Immigration and Naturalization Service with apparently no qualms about inviting foreign European labor to Nevada, despite his attacks against Basque immigrants. His response was in harmony with the views of ranchers that European labor was preferable to the closer-at-hand Mexican labor for these tasks. Despite his advocacy for cattle interests, he was evidently still responsive to the needs of his constituency. It was an immigrant labor lobbying effort that Senator Patrick McCarran would also take up decades later in his campaign to import Basque labor during the labor shortage in World War II and after.[67]

On October 20, Pittman sent a letter to Anthony Caminetti, commissioner general of Immigration, asking that restrictions on immigration be lifted to allow the recruitment of Romanian and Armenian immigrants.[68] Later, the office of the Department of Labor informed Pittman that immigrant labor from those eastern European countries could be recruited under the clause of the "skilled" laborers in the 1917 Immigration Act. But World War I concluded in the following year, November 1918, without any efforts to recruit sheepherders from Romania and Armenia.[69] In the meantime, an increasing number of Basque youths wanted to immigrate to the West and earn money in America, following in the steps of their older brothers and uncles who had immigrated earlier. At that time, according to the office of the

Table 4.2. Average daily wage and hours of stock-raising labor in Nevada

Occupation	Number of employees	Average wages per day	Average hours per day	Average pay for Nov. 1917
Cowboys	205	$1.76	9.6	$46.20
Camp tenders	129	$2.11	10.2	$63.29
Foremen	168	$2.61	9.9	$78.55
Sheepherders	774	$1.95	10.5	$56.73

Source: Second Biennial Report of the Commissioner of Labor of the State of Nevada, 1917–1918 (Carson City: State Printing Office, 1919), 42.

commissioner of Labor of Nevada, the average pay for a sheepherder per day was $1.95.[70]

Nevertheless, despite the requests for European labor in the sheepherding industry, Congress increasingly entertained proposals for additional restrictions on European immigration after World War I. It further tightened immigration rules in 1921 with the Emergency Quota Act that introduced a system of national quotas limiting annual immigration up to 3 percent of the number of each nationality's inhabitants existing in the United States based on the 1910 census. Because some Basques were in Spain and some in France, their respective populations were included in the quotas for those two countries. Between 1921 and 1924, the total number of immigrants permitted to enter annually under the quota remained the same for both Spain and France: the quota for Spain was set at 912 and for France at 5,729 immigrants.[71]

In May 1924, Congress passed a new immigration act, the Johnson-Reed Act, which further reduced the quota numbers allowing entrance of newcomers. Compared with the previous 1921 legislation, the 1924 Immigration Act was more stringent. The new immigration act reduced the quota number to 2 percent based on the 1890 census instead of the 1910 census, which reduced the numbers from eastern and southern European countries. In comparison with the French numbers, again, the immigration quota from Spain was sharply limited because there were fewer Spanish immigrants in the United States. The annual quota for Spain was set at 131 and for France at 3,954.[72] From 1924 until the passage of the 1965 immigration act, Basques generally had to find their places within these quota numbers.

By looking at the list of quota numbers set for European countries, it is clear how the act of 1924 favored the immigration from northern

Table 4.3. Annual quota numbers, 1921–1924

	Fiscal year 1921–1922		Fiscal year 1922–1923		Fiscal year 1923–1924	
Country	Quota	Number admitted	Quota	Number admitted	Quota	Number admitted
Spain	912	888	912	912	912	912
France	5,729	4,343	5,729	5,034	5,729	5,729

Source: Twelfth Annual Report of the Secretary of Labor. For the Fiscal Year Ended June 1924, (Washington, D.C.: Government Printing Office, 1924), 50.

Europe and largely discriminated against the immigration from southern and eastern Europe. In the case of the Basque Country, the quota number for France was higher than that applied to Spain. The provisions systematically reduced the opportunities for those Basques willing to immigrate from the provinces of Biscay, Guipúzcoa, Álava, and Navarre in Spain. In his book *Following the Grass*, American novelist Harry Sinclair Drago dealt with Basque sheepherders in northern Nevada and referred to the Pyrenees in their home country as "the great Basque barrier" protecting them from the outside world.[73] Of course, on one side of the Pyrenees were the Spanish-speaking Basques and on the other side French speakers. For its part, the American immigration quota law created another "Basque barrier"—limiting Spanish Basque immigration more severely than that from the French side of the Pyrenees.

The immigration act of 1924, historian Mae M. Ngai has argued, "differentiated Europeans according to nationality and ranked them in a hierarchy of desirability." The 1924 act set forth a new American immigrant labor structure in terms of whiteness during the mid-twentieth century. Congress invented the national origins quota system, in Mae Ngai's words, "that paradoxically upheld both the inviolate nature of racial bloodlines and the amalgamation of the descendents of European nationalities into a single white American race." And if the American whiteness construction was reinforced after the passage of this law, Ngai says, so was the idea of the "backwardness and unassimiability of the nonwhite races."[74] Between 1924 and 1965, as historian Matthew F. Jacobson has argued, white immigrants were blended together to form an amorphous group called Caucasians.[75]

At that time, like many other immigrants who left Europe in the late nineteenth and early twentieth centuries, some authors classified Basques as Caucasian, despite the many "mysteries" surrounding them.[76] In the early 1920s, "Basque" meant sometimes both "white" and "Caucasian" race in the American society.[77] On February 4, 1928, significantly, *The New York Times* included Basques in the list of "Caucasians and admitted as such with other aliens defined in the law as 'white.'" Whiteness enabled Basques to be seen as an increasingly exceptional "industrious" labor. As soon as Basque immigrants were labeled as Caucasians, their desirability in the host society increased. In this sense, American newspapers continued stating that Basques were a distinct and pure race in Europe, and further, a "Caucasian race."[78]

From the 1920s onward, although Basque immigrants were increasingly integrated into the local societies in the American West as white and assimilable subjects, they continued to suffer from sporadic racist assaults. When second-generation Basque Americans began attending school, they oftentimes encountered discrimination from English-speaking students. Juanita Olechea, a second-generation Basque-American woman, remembered being occasionally called "black Bascos" and "black Spaniards" at the junior high school of Gooding (Idaho) in the early 1930s.[79] There were undoubtedly many others who had similar experiences, such as Juanita Yribar or Reme Yturri in Idaho.[80] Also, racial discrimination intersected with religious discrimination, largely in the predominantly Protestant towns. Lucy Astorquia Osborne remembered being called "Black Basco" by the larger Methodist local community in Gooding, where it was not very popular to be Basque at the time.[81] Furthermore, social and economic success of some Basque families led to envy and this envy was oftentimes expressed in traditional negative attitudes, stereotypes, and insults. Some years later, for example, Mari Artecheverria Gabicagogeascoa remembered how some local people in Nampa (Idaho) called them and other Basques "Black Bascos" because they were jealous of some of the Basques' success and accomplishments.[82]

In the late 1920s, at the same time as Basques were integrating into the white mainstream, new Basque immigrants encountered increasing difficulties entering the United States. However, the decline in Basque immigration actually helped accelerate the integration process of the Basque-American community in the West as ranchers began to face difficulties employing Basque herders. Ranchers became convinced that Basque immigrants were better workers than others. Because of a growing scarcity of labor, Mexicans began to replace Basques as sheepherders. As I discuss later in chapter 6, the subsequent racialized othering of Mexicans in turn reinforced the notion that Basques represented an exceptional and anomalous racial stock in the scale of being "white," making them even more desirable. In the eyes of many western sheep owners, the Basque were a cohesive, cheap, and nonunionized labor immigrant group. This last factor would be particularly important during the thirties when workers' unions were losing strength in the United States.[83] Eventually, their reputation as diligent and loyal workers would make them preferable to other labor sources in the sheep industry of the American West.[84]

The Taylor Grazing Act and Ongoing Range Competition

As the interwar period progressed, tensions and conflicts in the public grazing lands of Nevada and other western states persisted, ultimately stirring congressional attention.[85] In June 1934, Congress passed the Taylor Grazing Act. The act was a significant step by the federal government toward the regulation of the remaining unclaimed western public domain lands. As implemented, it sought to organize grazing districts with decentralized administration by delegating powers of decision making about range matters to the local established graziers. Longtime sheep-grazing migration patterns overlapped state boundaries, but the new grazing districts partitioned these transhumance routes disregarding traditional patterns of migration and use. The new Taylor Grazing Act, although implementing local control, oftentimes disturbed customary patterns of sheep grazing. The new administrative system demanded much negotiation and explanations with ranchers.[86] The Taylor Act was the New Deal's response to the multiple complaints of the ranching community from Nevada and the entire West about the chaotic grazing conditions on the public domain exacerbated in Nevada by "tramp" sheepherders, largely associated with Basque immigrants. Of course, contrary to the popular conception, many Basques already owned lands and home properties, which was the requirement for holding grazing permits in national forests as would be the requirement in the new grazing districts.[87]

During the first half of the 1930s, the new federal grazing regime placed further burdens on the sheep industry by subjecting it to control by the new grazing boards in the districts oftentimes whose membership represented cattle interests. Furthermore, climate conditions brought drought, increasing costs for the purchase of feed to sustain flocks. This occurred during the economic downturn of the Great Depression that put additional strain on the sheep industry of Nevada. As weather conditions improved along with the economy, the sheep industry seemed to survive, but it was becoming clear as this decade progressed that the free and open range would be increasingly restricted with the threat of grazing fees.[88] Moreover, available immigrant labor, especially from the Basque Country, was nowhere at hand.

The outbreak of the Spanish Civil War (1936–1939) made things much more difficult for those ranchers desiring to recruit Basques. In the summer of 1937, the Fascist forces had conquered the Basque Country, pushing many Basques into exile. In many parts of the world, the Basque diaspora played an important role in helping their fellow countrymen to relocate to the host countries. For example, Basque immigrant communities were institutionally consolidated as never before all over Latin America.[89] In 1939, the hunger and political oppression that followed Franco's victory in the Spanish Civil War over the Republican forces led many Basques and others to flee Spain. Many resettled as refugees in countries open to their emigration. After France, Mexico became the second major country of destination for the Spanish refugees. It is estimated that around 20,000 Spanish exiles entered Mexico during the 1940s. Of the total number, more than 1,500 Basques from the provinces of Biscay, Guipúzcoa, Álava, and Navarre fled to Mexico. During and after the civil war, however, any Basque willing to escape to the United States with the pretext of working in sheep grazing or eager to work for any wage was subjected to the Spanish quotas still remaining in force.[90]

The United States, for its part, did not open its doors to Spanish refugees. President Franklin D. Roosevelt early expressed sympathy for the cause of the Spanish Second Republic. By the time of the civil war in 1936, nonetheless, the United States adopted a neutral position on the Spanish conflict in accordance with the Neutrality Acts of 1935 and 1936. In April 1937, while Germans and Italians helped to advance Franco's army, the United States reaffirmed its neutral stance after Congress by joint resolution adopted a stronger neutrality legislation declaring an embargo on arms sales to Spain. During the Spanish Civil War, FDR's friendly posture toward the Republican government confronted strong opposition from more conservative American elements and particularly the implacable hostility of many influential Catholics, who mobilized against the Spanish Republican government because of its anticlericalism. In 1937, significantly enough, American Catholic influence prevented a group of

Basque refugee children from entering the United States. During the last stage of the civil war, Roosevelt tried to send humanitarian aid to Spain with little success.[91]

By the late 1930s, the increasing regulation of the public grazing lands, the shortage of agricultural labor, and generally increasing production costs placed the sheep industry in a difficult situation as it faced the future. Many western woolgrowers blamed their deteriorating economic situation exclusively on federal regulation of rangelands. On this score, Senator Patrick McCarran, the second senator from Nevada alongside Pittman who had been elected in 1932 and who was a sporadic if not constant critic of the New Deal, was only too happy to come to the defense of Nevada's grazing community against federal grazing regulatory agencies. In the summer of 1938, McCarran protested vehemently about the negotiations of a trade agreement between the United States and Great Britain to remove the *ad valorem* duty on British import goods, which included removal of the wool tariff. Although McCarran's complaints fell on deaf ears, his efforts on several fronts at this time can be considered as a prelude to his subsequent fights in the years ahead on behalf of the sheep ranching community in Nevada extending into the postwar period, especially his attention to the Basque labor question.[92]

Summary

In the early decades of the twentieth century, the Basques gained a reputation as a desirable immigrant community. The Great Basin provided the geographical space where Basque immigrants progressively adjusted themselves in their own ways by both maintaining their culture and accommodating themselves to the larger society. Despite Basque retention of a distinct identity and culture, they became part of the dominant and more complex American society. By the 1910s, a Basque community was consolidated in Nevada. Their affiliation with the Catholic Church and the many traditional landmark Basque hotels in Nevada strengthened ties and relations outside their community's borders. More importantly, many early Basque immigrants had purchased land, acquired water rights, and established prominent sheep outfits in Nevada. All these factors indicated that a great percentage of the Basque immigrant community had come to stay.

Restrictive immigration legislation that occurred during and following World War I severely curtailed Basque immigration to the United

States. The Immigration Act of 1917 worked against subversive aliens and Asian immigration, affecting Basque immigration only slightly. It did place an emphasis upon "skilled" workers, if the necessity of employing foreign labor could be demonstrated because of the scarcity of such labor in the country. However, the subsequent quota acts of 1921 and especially 1924 severely affected Basque immigration, primarily because of the quota applied to all of Spain. The full effects of the 1924 immigration law were not felt until a decade later, when in the late 1930s an important Basque labor shortage was apparent in the open-range sheep industry in Nevada and the entire West. During these interwar years, Basque immigrants became racially accepted laborers, if not prized laborers, resulting from their so-called racial characteristics and, of course, their whiteness.

In the late 1930s, among other issues, the recruitment of Basque sheepherders became a pressing issue among western woolgrowers. The restrictive quota numbers applied to Spain reduced the opportunities for Basque job seekers in the western sheep industry. During the Spanish Civil War, in addition, many Basques were exiled from their homeland because of political oppression. Although many were willing to immigrate to the United States, U.S. immigration restrictions barred the Basque exiles. Now sheep ranchers foresaw a chronic labor shortage in the sheep industry that moved them to petition Congress in the coming years to permit additional Basque immigration. To obtain special legislation, they touted Basques as "skilled" workers and emphasized their alleged racial and cultural talents for sheep grazing as justification to grant them permanent "skilled" labor immigrant permits. The next chapter examines these issues in depth, specifically the Basque immigrant labor shortage in the western sheep industry during World War II and the important role that Senator Patrick McCarran played defending vigorously the recruitment of Basque immigrants despite the legal and international impediments.

Notes

1. Birth Certificate, Angela María Odriozola y Larrucea, September 15, 1903, certificate no. 35321, recorded in the book no. 8, p.50, Gernika-Lumo Court, Biscay, Spain, Angela Aguirre Papers, Aguirre Family Archive, Elko, Nevada; Baptismal Certificate, Angela María Manuela de Odriozola y Larrucea, September 16, 1903, recorded in the book no. 6, p.204, Parish Church St. Pedro, Gernika, Biscay, Spain,

Angela Aguirre Papers; Interview, Angela Aguirre, August 27, 1993, transcript, Oral History Collection, Northeastern Nevada Museum, Elko, Nevada.

2. Medical Certificate, Angela Odriozola y Larrucea, June 7, 1922, certificate no. 5137, College of Physicians of Biscay, Angela Aguirre Papers; Labor Certificate, Angela Maria Manuela Odriozola y Larrucea, June 7, 1922, Gernika-Lumo Court, Angela Aguirre Papers; Parental Consent, Pedro Odriozola y Guerricaechevarria, June 8, 1922, Gernika-Lumo Court, Angela Aguirre Papers; Passport, Angela María Manuela de Odriozola y Larrucea, June 9, 1922, Spanish passport no. 1914, Angela Aguirre Papers.
3. Passport, Angela María Manuela de Odriozola y Larrucea, June 9, 1922; Aguirre Oral History, Northeastern Nevada Museum; Ronald H. Bayor, *Encountering Ellis Island: How European Immigrants Entered America* (Baltimore, Md.: Johns Hopkins University Press, 2014), 50.
4. Aguirre Oral History, Northeastern Nevada Museum; Marriage License and Certificate, Guy Aguirre and Angela Odriozola, Nov. 3, 1923, license no. 35321, recorded in the book no. 4, p.23, Elko County Courthouse, Elko, Nevada; as also noted in my article in the *Elko Daily Free Press*, January 20, 2015.
5. John Higham, *Strangers in the Land: Patterns of American Nativism, 1860–1925* (New Brunswick, N.J.: Rutgers University Press, 1955), 87.
6. Arrizabalaga, "Statistical Study of Basque Immigration," 53–54.
7. *Egunaria edo Almanaka: Eliza-ofizioetako aurkhi-bidea* (Bayonne: Lasserre, 1911), 7.
8. *Eskualduna*, May 27, 1910, 2.
9. "Ameriketarat," *Eskualduna*, May 13, 1910, 3.
10. Shepperson, *Restless Strangers*, 14; Echeverria, *Home Away from Home*, 36–74, 134–65; "The New Star Hotel," *Daily Free Press*, December 16, 1910.
11. *Daily Free Press*, November 14, 1911, 4.
12. Echeverria, *Home Away from Home*, 43–45.
13. Ibid., 218–30; Lourinda R. Wines, "The Long Journey," 1973, unpublished manuscript, Special Collections, Library University of Nevada, Reno, Nevada.
14. Jeronima Echeverria, "*Euskaldun Andreak*: Basque Women as Hard Workers, *Hoteleras*, and Matriarchs," in *Writing the Range: Race, Class, and Culture in the Women's West*, eds. Elizabeth Jameson and Susan Armitage (Norman: University of Oklahoma Press, 1997), 298–310; Mary Jane Etcheverry, unpublished autobiography, Manuscript Collection, Eureka Sentinel Museum, Eureka, Nevada; Anne M. Butler, *Daughters of Joy, Sisters of Misery: Prostitutes in the American West, 1865–1890* (Urbana and Chicago: University of Illinois Press, 1985), ix; on the role of Basque women in the livestock industry in Nevada, see also Evelyne L. Pickett, "Women in the Empty Quarter: A Study of Changes and Challenges as Related to Women's Experiences in the Nevada Ranching Industry from the Mid-Nineteenth Century to the Late Twentieth" (master's thesis, University of Nevada, Reno, 1988).
15. On the Irish in the Nevada's mining industry, see Ronald M. James, *The Roar and the Silence: A History of Virginia City and the Comstock Lode* (Reno: University of Nevada Press, 1998).

16. James S. Olson, "Pioneer Catholicism in Eastern and Southern Nevada, 1864–1931," *Nevada Historical Society Quarterly* 26, no. 3 (Fall 1983): 159–71; Ferenc M. Szasz, *Religion in the Modern American West* (Tucson: University of Arizona Press, 2000), 33.
17. Ronald M. James, "Erin's Daughters on the Comstock: Building Community," in *Comstock Women: The Making of a Mining Community*, eds. Ronald M. James and C. Elizabeth Raymond (Reno: University of Nevada, Reno, 1998), 246–62; Kevin Rafferty, "Catholics in Nevada," in *Community in the American West*, ed. Stephen Tchudi (Reno and Las Vegas: Nevada Humanities Committee, 1999), 201–16; historian Steven Avella has suggested that the Catholic Church in Sacramento, while confronting the material desire in the American West, was integral to the life and identity of different immigrant groups existing in the city. Steven M. Avella, *Sacramento and the Catholic Church: Shaping a Capital City* (Reno: University of Nevada Press, 2008), 100–31; "Bishop Allows Lenten Union," *Nevada State Journal*, March 8, 1911, 8; "Catholic Services," *Reno Evening Gazette*, April 23, 1910; "Priest Visits Basque Ranches," *Reno Evening Gazette*, March 29, 1932, 12; "Speaking to the Basque Population," *Nevada State Journal*, March 30, 1932, 8.
18. Douglass and Bilbao, *Amerikanuak*, 355–59; Hovey, *Portraits of Basques in the New World*, 78.
19. "Father Thomas Tubman Expresses Himself Relative to State's Basque Colony," *Reno Evening Gazette*, April 1, 1911, 2.
20. Father Thomas Tubman enjoyed good personal relationships with the Basques of Nevada. He had conducted many baptisms, weddings, and funerals for the Basque immigrant population. Sometimes, the local press reported Basque weddings. Many articles noted that the married couples were prominent members of the local community. Similar events and reports occurred in other Nevada towns such as Elko. Some few examples are as follows: May 24, 1910: Isidoro Sara and Marie Arbans; August 9, 1910: Juan Uhalde and Jeanne Marie Borda; February 11, 1911: Jose Oznarez and Catalina Etulian; March 11, 1911: Pedro Urguiza and Felicia Villanueva, "Prominent Basques United in Marriage," *Reno Evening Gazette*, May 25, 1910; "Members of Basque Colony Married," *Reno Evening Gazette*, August 10, 1910, 2; "Father Tubman Weds Prominent Basques," *Nevada State Journal*, February 12, 1911, 8; "Basque Colony Couple United," *Nevada State Journal*, March 13, 1911, 6; "Popular Spanish Basque Couple Wed," *Daily Free Press*, August 12, 1915.
21. "Desirable Immigrants," *Reno Evening Gazette*, March 29, 1911, 4.
22. Ibid.
23. Ibid.
24. Resurrección María de Azkue, *Ardi Galdua* (Bilbao: Corazón de Jesús, 1918).
25. Lane, *Anglo-American Contributions to Basque Studies*, 33–36.
26. *Carson Morning Appeal*, May 4, 1898, 3.
27. *California Cultivator and Livestock and Dairy Journal* 13 (1899): 97.
28. Hays, *Conservation and the Gospel of Efficiency*, 54.
29. Austin, *The Flock*, 62. Mary Austin understood that these immigrants who brought with them their cultural traditions and ways of life were socioculturally preadapted to sheepherding in the arid West. Contrary to this assumption, Basque

immigrants were not preadapted and qualified for extensive sheep grazing, but like many others had to learn and adjust to this new job in the American West.

30. Lane, *Anglo-American Contributions to Basque Studies*, 35.
31. Richard W. Etulain, "The Basques in Western American Literature," in *Anglo-American Contributions to Basque Studies*, 8.
32. Interview, Justa Yturri, August 2, 1975, transcript, Oral Histories, Basque Museum and Cultural Center, Boise, Idaho; Bieter and Bieter, *Enduring Legacy*, 156; Powers, *American Merino*, 249; Harry Sinclair Drago, *Following the Grass* (New York: The Macaulay Company, 1924), 21; Echeverria, *Home Away from Home*, 32.
33. Interview, Timotea Yraguen Echanis, March 16, 2002, transcript, Oral Histories, Basque Museum and Cultural Center, Boise, Idaho.
34. "Basque Language Balks Inspectors," *New York Times*, March 21, 1911, 9; "The Queerest Things in the World," *Boston Post*, December 11, 1917.
35. "Exiled Basques in California Go Mad Tending Their Flocks," *Washington Post*, April 7, 1907, 9.
36. Ibid.
37. See "foreignization" in Robert E. Rhoades, "Foreign Labor and German Industrial Capitalism, 1871–1978: The Evolution of a Migratory System," *American Ethnologist* 5, no. 3 (August 1978): 553–73. On the ideological basis of American nativism, see Higham, *Strangers in the Land*.
38. Fred L. Israel, *Nevada's Key Pittman* (Lincoln: University of Nebraska Press, 1963), 19–34.
39. "Pittman's Sneer," *Reno Evening Gazette*, October 7, 1910, 4.
40. Ibid.
41. "The Statement of Senator Pittman of Nevada," *National Wool Grower* 3, no. 11 (November 1913): 21.
42. Ibid.
43. Ibid.
44. Israel, *Nevada's Key Pittman*, 26; "Statement of Senator Pittman," *National Wool Grower*, 21–22.
45. "Speaks a Good Word for Basques," *Silver State*, July 24, 1913, 1; the letter was reprinted in the Elko newspaper: "Some of Our Best Citizens Are Basques," *Daily Free Press*, July 29, 1913.
46. "Speech of Hon. Key Pittman," Cong. Rec., 63rd Cong., 1st Sess., (August 21, 1913), Key Pittman Papers, Nevada Historical Society, Reno, Nevada.
47. Ibid.
48. Ibid.
49. Ibid.
50. Ibid.
51. "Statement of Senator Pittman," *National Wool Grower*, 23.
52. Ibid., 23.
53. Ibid., 23–24.
54. "Sir:

Your esteemed editorial on "Senator Pittman's Estimate" of the sheep industry, and the men that conduct it in Nevada, fell like a bomb into our camp. We felt happy over the able way in which you and Mr. [Jerry] Sheehan of the Winnemucca First National bank, have disposed of the misrepresentation of the industry by that "misrepresentative" of ours, and I shall not dwell on that phase of his testimony.

His description of the sheepmen in Nevada is in keeping with his estimate of the industry itself, and has proved a veritable bolt from a clear sky.

It is concerning this feature of his testimony that I mean to enter my indignant protest, before the people of the state of Nevada, and before the ruling powers in Washington.... Nine-tenths of them [Basques] own their land and their homes, dependence being considered the lowest condition of social life, and none being ever excused from work, that has a mind to think and a hand to wield...

Have they then mastered masterminds in statesmanship and war for 30 centuries, with intelligence? But how could we expect an ignoramus to be acquainted with the great men of the past, when he displays mulish ignorance of his own constituents and their interest, in his own day and generation, right at his own door...

Who have dotted the California coast with successfully managed and prosperous ranches for the last 50 years, but some of these sturdy Pyrennean mountaineers? What did the sheep industry amount to in Nevada, until they came, and made this industry the biggest in the state, notwithstanding the senator's assertion to the contrary?

It does not take book lore to create and keep up the sheep industry, but it does take intelligence, dogged tenacity, sobriety and business acumen, such as all university professors and senators do not always have, to push it to a successful issue.

Note well, there are scores of Basques, young and old, who speak and write fluently in four or five languages..." The letter was reprinted in the Elko newspaper: "Prominent Member of Basque Race Takes Nevada's Senator to Task for Remarks Made," *Daily Free Press*, October 21, 1914.

55. "Attack by Pittman on Basques Is Challenged," *Reno Evening Gazette*, October 30, 1916, 5; later, in 1917, Sol Silen published his book *La Historia de los Vascongados en el Oeste de los Estados Unidos*.
56. "Speech of Hon. Key Pittman," Nevada Historical Society.
57. William D. Rowley, "Historical Considerations in the Development of Range Science," in *Forest and Wildlife Science in America: A History*, ed. Harold K. Steen (Durham: Forest History Society, 1999), 247; Rowley, *U.S. Forest Service Grazing*, 67–73; Steen, *The U.S. Forest Service*, 163. Nevada Historical Society, Reno, Nevada.
58. *Statutes of the State of Nevada, Thirty-First Session of the Legislature 1923* (Carson City: State Printing Office, 1923), 25–31; "Two Are Shot in Austin Fight," *Reno Evening Gazette*, July 10, 1924, 3.
59. D. Staunton to Key Pittman, March 25, 1914, Box 64, File Department of Agriculture; Chester A. Laing to Key Pittman, December 8, 1916, Box 112, File Stock Raising Homestead Bill; T.J. Bell to Key Pittman, April 23, 1917, Box 124, File Miscellaneous Correspondence; Key Pittman to T.J. Bell, June 4, 1917, Box 124, File Miscellaneous Correspondence; Advisory Board of the North Fork Livestock Association to C.E. Favre, November 15, 1917, Box 64, File Forest Service; Advisory Board of the North Fork Livestock Association to Key Pittman,

February 23, 1918, Box 64, File Forest Service; J.H. Clemons to the Forest Service office in Ogden, Utah, March 20, 1918, Box 64, File Forest Service; Timothy Murphy to Key Pittman, April 26, 1918, Box 64, File Department of Agriculture; Key Pittman to James Dysart, June 2, 1923, Box 104, File Grazing Reserves; Key Pittman to William B. Greely, February 21, 1924, Box 108, File Nevada National Forests, Key Pittman Papers, Library of Congress, Washington, D.C.

60. Key Pittman to William B. Greely, February 21, 1924, Pittman Papers.
61. Leon F. Kneipp to Key Pittman, April 12, 1924, Box 108, File National Forests, Pittman Papers.
62. Hays, *Conservation and the Gospel of Efficiency*, 266; Leon F. Kneipp to Key Pittman, April 12, 1924, Box 108, File National Forests, Pittman Papers.
63. F.N. Smith to Key Pittman, April 7, 1924, Box 108, File Nevada National Forests, Pittman Papers.
64. Fitzgerald, *The Face of the Nation*, 127–30; Immigration Act of 1917, H.R. 10384, 64th Cong., 2nd Sess., Pub. L. 301, 39 Stat. 874, February 5, 1917, in *Statutes of the United States of America, Passed at the Second Session of the Sixty-Fourth Congress, 1916–1917, Part I: Public Acts and Resolutions* (Washington, D.C.: Government Printing Office, 1917), 874–98.
65. Immigration Act of 1917, 877.
66. William A. Keddie to Key Pittman, September 28, 1917, Williams Family Archive, Fallon, Nevada.
67. Key Pittman to William A. Keddie, October 8, 1917, Williams Family Archive.
68. Key Pittman to Anthony Caminetti, October 20, 1917, Williams Family Archive.
69. E.J. Trenwith to William A. Keddie, October 30, 1917, Williams Family Archive.
70. *Second Biennial Report of the Commissioner of Labor of the State of Nevada, 1917–1918* (Carson City: State Printing Office, 1919), 42.
71. Ngai, *Impossible Subjects*, 21–22; *Twelfth Annual Report of the Secretary of Labor. For the Fiscal Year Ended June 1924* (Washington, D.C.: Government Printing Office, 1924), 50–51.
72. *Twelfth Annual Report of the Secretary of Labor. For the Fiscal Year Ended June 1924*, 51–55.
73. Fitzgerald, *The Face of the Nation*, 150–52; Drago, *Following the Grass*, 11.
74. Ngai, *Impossible Subjects*, 24, 37, 25.
75. Jacobson, *Whiteness of a Different Color*, 91–96.
76. *Evening Independent*, October 20, 1922; *Lethbridge Herald*, October 20, 1922, 4; *Des Moines Capital*, October 20, 1922, 8.
77. *Nevada State Journal*, February 12, 1924, 4.
78. *New York Times*, February 4, 1928, 26; Eleanor Davis, "The Basques in Malheur County," *Commonwealth Review* 9, no. 2 (April 1927): 55.
79. Interview, Juanita Olechea, September 4, 2001, transcript, Oral Histories, Basque Museum and Cultural Center, Boise, Idaho.
80. Interview, Juanita Yribar, September 14, 2001, transcript, Oral Histories, Basque Museum and Cultural Center, Boise, Idaho; Interview, Reme Yturri, March 16, 2002, transcript, Oral Histories, Basque Museum & Cultural Center, Boise, Idaho.

81. Interview, Lucy Astorquia Osborne, June 26, 2002, transcript, Oral Histories, Basque Museum and Cultural Center, Boise, Idaho.
82. Interview, Mari Artecheverria Gabicagogeascoa, January 16, 1976, transcript, Oral Histories, Basque Museum and Cultural Center, Boise, Idaho.
83. Aurora Bosch, "Estados Unidos en los años treinta: ¿Un socialismo imposible?" *Historia Social*, no. 11 (Fall 1991): 41–42.
84. Saitua, "Distilling Spirits," 64.
85. In 1932, a report by agricultural specialist Elmer Otis Wooton revealed that one of the most serious problems affecting Nevada's public rangelands was "the encroachments of 'tramp' sheep" and noted that all previous legislation upon this subject was "futile." In his report, Wooton made it plain to Congress that in Nevada there existed an urgent and great necessity for federal legislation concerning the use of the public grazing lands. Wooton, *The Public Domain of Nevada and Factors Affecting Its Use*, 47, 3, 44–47.
86. Foss, *Politics and Grass*, 61–72, 92–116; Wesley Calef, *Private Grazing and Public Lands: Studies of the Local Management of the Taylor Act* (Chicago: University of Chicago Press, 1960), 52–57; H.R. Hochmuth, Earl R. Franklin, and Marion Clawson, *Sheep Migration in the Intermountain Region*, USDA Circ. 624 (Washington, D.C.: Government Printing Office, Jan. 1942), 3–4, 29–30, 40–41; "Utmost Benefits to Be Sought by Grazing Plan," *Reno Evening Gazette*, September 25, 1934, 14.
87. Calef, *Private Grazing and Public Lands*, 56–57; Hatfield, "We Were Not Tramp Sheepmen," 313–25, 432–53; Kenneth B. Platt, "Taylor Grazing Law Features Discussed," *Nevada State Journal*, February 29, 1940, 7; historian Karen Merrill notes "slippages" in meaning, conversations, and in the relations between the grazing community and federal administrators over the decades. Karen R. Merrill, *Public Lands and Political Meaning: Ranchers, the Government, and the Property between Them* (Berkeley: University of California Press, 2002), 2, 205–10.
88. Hochmuth, Franklin, and Clawson, *Sheep Migration in the Intermountain Region*, USDA Circ. 624 (Washington, D.C.: Government Printing Office, Jan. 1942): 1–5; Charles E. Flemming and C.A. Brennen, *Studies of the Range Sheep Business in Nevada*, Bulletin no. 145 (Reno: University of Nevada, Reno, Aug. 1937), 6, 12–13.
89. Gloria P. Totoricagüena, *Identity, Culture, and Politics in the Basque Diaspora* (Reno: University of Nevada Press, 2004), 68–73; J. Giralt C., ed., *Los Vascos en América. Tomo I: Chile, Argentina y Uruguay* (Santiago de Chile: Cathalonia, 1937), 7; Matthew F. Jacobson notes that "collective emigration nourished a political culture based on ideas of injury and displacement." Matthew F. Jacobson, *Special Sorrows: The Diasporic Imagination of Irish, Polish, and Jewish Immigrants in the United States* (Cambridge, Mass.: Harvard University Press, 1995), 2; see also: Koldo San Sebastián, *El exilio vasco en América: 1936/1946–Acción del Gobierno* (San Sebastián: Txertoa, 1988); Xabier Irujo, *Expelled from the Motherland: The Government of President Jose Antonio Agirre in Exile (1937–1960)* (Reno, Nev.: Center for Basque Studies Press, 2012); David Mota Zurdo, *Un sueño americano: El Gobierno Vasco en el exilio y Estados Unidos (1937–1979)* (Oñati: IVAP, 2016).

90. Manuel Ortuño Martínez, "Fin de la guerra y largo exilio," in *Socialismo y guerra civil*, ed. Santos Juliá Díaz (Madrid: Fundación Pablo Iglesias, 1987), 381–95; Clara Eugenia Lida, comp., *Una inmigración privilegiada: comerciantes, empresarios y profesionales españoles en México en los siglos XIX y XX* (Madrid: Alianza, 1994), 220–22.
91. Aurora Bosch, "Entre la democracia y la neutralidad: Estados Unidos ante la Guerra Civil española," *Ayer* 90, no. 2 (2013): 167–87; Dominic Tierney, *FDR and the Spanish Civil War: Neutrality and Commitment in the Struggle That Divided America* (Durham and London: Duke University Press, 2007), 63, 121–26, 135–36, 157; Peter N. Carroll and James D. Fernández, eds., *Facing Fascism: New York and the Spanish Civil War* (New York: Museum of the City of New York and New York University Press, 2007), 95–97; Soledad Fox Maura, "Miradas Opuestas: la Casa Blanca y la opinión pública norteamericana ante la guerra de España," *Circumstancia* 19 (Mayo 2009); the relations of the FDR Administration with the Second Spanish Republic have been comprehensively studied in the book: Aurora Bosch, *Miedo a la democracia: Estados Unidos ante la Segunda República y la guerra civil española* (Barcelona: Crítica, 2012).
92. "Protest Trade Treaties," *Idaho Wool Growers Bulletin* 18, no. 14 (July 6, 1938): 1; Patrick McCarran to John Dangberg, May 4, 1942, Collection no. 13, Box 38, Dangberg Correspondence 1942, Patrick McCarran Papers, Nevada Historical Society, Reno, Nevada.

PART III

The Making of a Good Sheepherder

5

"Grasping at a Straw"

The Basque Labor Shortage in the Nevada and Western Sheep Industry during the Second World War

On June 5, 1940, the Idaho Wool Growers' Association, addressing the problems confronting western sheep operators, published in its bulletin an article by Bill Johnston, a student at the University of Idaho, entitled: "In Defense of Sheepherders." It noted that in Idaho, as elsewhere in the West, considered the "grazing ground of some of the nation's finest herds of sheep, the phrase, 'crazy as a sheepherder,' is part of the language." Johnston noted that sheepherders were socially discredited workingmen and were seen as an underclass in agricultural labor. According to him, the labor of sheepherding was something "unusual," but not "insane." Mockers ridiculed the herders as crazy men, Johnston explained, because they spent most of the year "roaming through huge mountain ranges at the rear end of a band of squalling sheep." Replying to those scoffers who laughed at sheepherders' work, Johnston raised the following question: "But is a scholar crazy, too, because he forsakes humanity to delve into books? If anything, the sheepherder has the advantage. He doesn't go blind so quickly." Johnston continued:

> If you say the sheepherder is a coward and runs away from life, you have never seen the angry lightning slashing down on barren ridges where an insignificant man stands alone with his scared, bewildered sheep. You have never seen the mass and power of a hungry grizzly. Perhaps, he is a moral coward, but who of us is not? Who dares to look at life carefully and steadily and never run away? Are you even sure the sheepman misses life because he shuns humanity? How long has it been, you busy hustlers, since you have seen a cloud?[1]

Johnston argued that the sheepherder's dirty, tedious, and lonely life was one of ambition and achievement. He concluded by saying: "Scorn not, then, 'the crazy sheepherder.' Perhaps he has a wisdom better than our own." Among the woolgrowers this view might have been readily accepted because their business profits depended on the sheepherders. Nevertheless, in 1940 when Johnston's article was published, these positive words would go largely unnoticed by the public and a generally negative view of sheepherders prevailed. But attitudes toward sheepherders would evolve during World War II when a serious labor shortage began to affect the industry.[2]

As the United States became more involved in the war, demand dramatically increased for agricultural products. Nevada's sheep industry was no exception. Consequently, sheep ranchers needed more sheepherders and demanded from their representatives in Congress that legal restrictions on Basque immigrants be lifted. According to them, the depletion of Basque labor to work on the ranges threatened to obstruct sheep production. They organized themselves to find solutions. In this fight for Basque sheepherders, Nevada's U.S. senator Patrick McCarran was instrumental in helping woolgrowers to "import" a Basque immigrant labor supply.[3] Although it was not the only issue facing the sheep industry in the West, it drew great attention.

Pat McCarran gave a high priority to the sheepherder question among other issues affecting the sheep industry of Nevada and the West. McCarran focused his attention on numerous issues surrounding the recruitment of Basques, putting in place the necessary machinery in order to allow an increasing Basque immigration in the United States. However, the 1920s' restrictive measures and the serious international crisis of the Second World War made difficult any effort to open the doors for more Basque immigrants. The present chapter analyzes how a group of sheep ranchers from western Nevada united to respond to the shortage of Basque labor. Between the spring season of 1942 and January 1943, under the patronage of Senator McCarran, these Nevada woolgrowers organized themselves to push government officials to allow the recruitment of further Basque immigrant workers.

The extensive correspondence among important industry figures and the broader political advocacy that occurred during this time also further underlined how forcefully the industry had begun to insist on the notion of Basque immigrants being bestowed by certain ethnic and racial attributes, particularly in comparison with other ethnic or national groups. This time, however, it was not in order to discriminate against Basque immigrants as the cattle ranchers had earlier attempted, but instead to claim that the Basques were bestowed with unique traits needed for the maintenance of the entire sheepherding industry.

Pat McCarran, John Dangberg, and Basque Labor in the Sheep Industry

To develop an understanding of how Pat McCarran and local ranchers worked together on the issue of the Basque labor during World War II, it's necessary to take into account the so-called "McCarran Machine" in Nevada politics and the overall picture of the sheep industry in the Great Basin and the West. At the time of his 1938 bid for reelection to the Senate, Pat McCarran was a popular Nevada politician. He was praised from many sectors, particularly from rural areas of Nevada, as the man who had defended the interests of the common people. On August 23, 1938, the Nevada edition of *Labor* gave prominence to McCarran's role in the development of Nevada and called for his reelection to the Senate. The *Labor* declared that, "interests' fear him [McCarran], because he's always on [the] people's side." Eventually, McCarran won reelection and increasingly gained notoriety within the Democratic Party. In 1940, after the death of Senator Key Pittman, McCarran became the senior senator from Nevada, thus beginning the McCarran era in Nevada politics.[4]

In the absence of Pittman, McCarran had the political power in the Senate to bring the Nevada Democratic Party under his influence. He formed the McCarran machine to dominate state politics. The "lone wolf," as historian Alfred Steinberg called him, created a network between Nevada and Washington, D.C., based upon a patronage system of personal favors and rewards for his Nevada friends. The national press frequently criticized Pat McCarran's political methods.[5] McCarran was deeply committed to the primary sectors of Nevada's economy. The senator focused a great deal of his energy in the upper house on promoting the two major industries of Nevada throughout the first half of the twentieth century: mining and stock agriculture. He strongly believed that the progress of the state's economy depended upon the health of these sectors. In addition, western stockmen were a considerable lobby and they influenced McCarran in determining policy.[6]

McCarran aligned himself squarely with the livestock industry. This tenacious and ambitious politician, influenced perhaps by his rural origins in Nevada, challenged any legislation that threatened directly his cattlemen and sheepmen constituency. McCarran was primarily interested in reducing the more active federal presence in land management. After the passage of the 1934 Taylor Grazing Act, the burden of federal fees (especially after 1936 when the first grazing fee was imposed) and other regulatory measures over the public domain raised and aggravated the unrest among stock operators. Responding to his constituency, the Nevada senator used his position as a member of the Senate Committee on Public Lands and Surveys to oppose the expansion of federal bureaucracy. McCarran, however, defended a new grazing system to determine priority rights for adjudication of public domain to private users. McCarran introduced Resolution 241 which gave the power to conduct hearings about the federal lands administration. Eventually, the Senate approved it in May 1940 and the Senate Committee on Public Lands and Surveys conducted hearings and even challenged the constitutionality of the Taylor Grazing Act.[7]

Similar to the situation during World War I, the U.S. involvement in World War II set the pattern of the economy of the American West. In general terms, the war became a catalyst for western economic development. In the wartime bonanza, industrial metal mining grew intensively, such as the extraction of copper, tungsten, zinc, and lead. Although it grew less than the mining industry, the state's agricultural income increased, too. During the war, the international economy further contributed to increased expectations among western woolgrowers and served to mobilize them in common causes.[8] In 1942, as an illustrative example, an advertisement for Purina Mills (the animal food company) made direct reference to the war effort and displayed a strong patriotic message to market its product of supplemental sheep fodder: "You Do the Fighting and We'll Produce the Lambs and the Wool."[9]

Despite the expansion of the Nevada economy during the war, and contrary to sheepmen's expectations, the early wartime economy proved frustrating for the sheep industry. The high demand for meat produced a moderate increase in the western livestock production, but the results were not satisfactory for stock operators. Unlike in the days of World War I, this time sheepmen found more obstacles to produce wool and mutton at a low price. By the mid-twentieth century, changes in land administration, shifts in agriculture economics affecting the demand for sheep, and the changing global market and food consumption patterns had all affected the sheep industry.[10] By the late 1930s, governmental regulation in public lands had notably altered the traditional sheep breeding and management. The Forest Service's wartime policies—in contrast to the previous experience of World War I—were more restrictive, and despite the high demand for the use of the range, the Forest Service continued to protect range resources. In the interwar period, moreover, improvements in industrial food refrigeration caused diminished consumption of lamb meat and led to a more diversified diet. In the late nineteenth century, because of their small size and the lack of refrigeration apparatus, sheep was a marketable and easily consumable meat. Nevertheless, the food preservation techniques improved in the twentieth century, the market diversified, and dietary customs changed, lamb and mutton began to lose their place in the national and international economy. But in addition to these many issues facing the industry, the war underlined another major problem: the shortage of cheap and effective immigrant agricultural labor.[11]

Recruiting Basque Sheepherders

The growing labor shortages, which occurred across the United States during the war, left the majority of sheepmen in a desperate situation, as they were unable to find dependable immigrant workers. In the late 1930s, the number of sheep workers, or to be more precise, the number of people willing and able to do the work, had already declined. But as the war advanced in the 1940s, uncertainty and problems regarding sheep labor emerged in Nevada and other western states. Simply put, there was too much stock with too few herders available in the West, and the effects of the national immigration restrictions in the 1920s were being significantly felt two decades later when wartime demanded more labor.

By that time, the axiom that Basque immigrants were the most qualified sheepherders had already become a cliché among western woolgrowers, and, to the maximum extent possible, ranchers gave priority to employing only Basques. As a result, despite the fact that the national origin-based quota system greatly reduced the number of poor immigrants arriving from Spain, sheep operators continued to recruit Basques to come to the United States, and some Basques would even come to the United States either illegally or on tourist visas. However, this situation inevitably posed problems.

Among those who came to work on a temporary visa were Serafin Larrabaster, Francisco Munitis, and Agustin Lozano who, in the summer of 1943, left Franco's Spain and immigrated to the United States. At that time, they would have expected to earn more than $100 a month working as sheepherders in the West (the average monthly wage paid to sheep workers was $120). Before their departure, the Yakima Sheep Company (one of the largest sheep operations in the Pacific Northwest) had already employed them without providing detailed information of the terms of the contract, working conditions, or immigration requirements. On August 17, 1943, they arrived at New Orleans onboard the steamship *Magallanes*. On their arrival, however, immigration authorities refused them entry to the United States because they had arrived on tourist visas valid for only one year, instead of the quota visa required from Spanish immigrants. But the three Basques did not know that the visa only permitted them to reside for one year, nor that after the authorized period, their employer intended to keep them undocumented until legalizing their residence status. At that time, in the post–Spanish Civil War years, any Basque immigrant coming from Spain entering on a tourist visa was automatically considered an exile that would likely intend to overstay his one-year permit. In this particular case, the immigration authorities permitted the three Basques to land and remain temporarily (in the depot for Algiers immigrants) in the city until their legal situation was clarified. In order to ensure that these immigrants would not overstay their temporary visa, the Immigration and Naturalization Service requested that the Yakima Sheep Company furnish a bond as a guarantee of these three Basques' passage back to Spain after their entry-visa expired. Although Franco's regime would initially try to block the emigration of these Basque workers, the workers would eventually reach their final destination. Later on, the Yakima Sheep Company noted how hard it had been to import the workers: "The Spanish government gave their permission to recruit these men, but it was not easy."[12]

In general, two main factors affected the Basque labor shortage: first, many of those Basques who had entered the United States from 1880 to 1924 retired, went back to Europe, or died; second, many of the first generation of Basque herders had diversified their economic interests. Indeed, since the wartime economy had opened up many new employment opportunities, sheep grazing became a less desirable occupation for Basque immigrants. Some Basque immigrants went into the military or abandoned sheepherding to enter jobs where they could find less tedious work, better conditions, and better pay. The departure of this workforce, and the impossibility of employing additional Basques from the Old Country put ranchers into a difficult position as they sought a new labor supply.[13]

Consequently, the ranchers organized themselves into associations to increase their political strength and advocated for the removal of the restrictions on Basque immigrants. The Basque labor shortage became a central issue within the industry in the wartime and postwar period, even leading woolgrowers to find common ground on other problems affecting the sheep industry. McCarran's work on behalf of Nevada ranchers caused one later political scientist, R. McGreggor Cawley, to comment that the senator was an early *sagebrush rebel*. As McCarran acquired attention for his stands against the Grazing Service, the ranching community applauded his efforts and saw him as a possible ally in their search to relax immigration laws for Basque sheepherder immigrants.[14]

As a national advocate for the sheepmen and the Basque immigrant population, McCarran was already well known for his strong sympathy toward the Basques. He knew very well the business and the work with the sheep itself from his own personal experience. He had spent his childhood working on his family sheep operation with a home ranch located in Washoe and Storey Counties, twelve miles east of Sparks, Nevada. During his early life, McCarran interacted with the Basque sheepmen, their communities, and the important Laxalt family, one of the most distinguished Basque-American families in Nevada. In the late 1920s, Paul Laxalt (governor of Nevada 1966–1970) recalled that when his mother Therese operated the French Hotel in Carson City, McCarran was among the most regular customers. McCarran was, in Laxalt's words, "like a deity in our household." Furthermore, John Laxalt, one of Paul's brothers, became a "McCarran boy" and became part of McCarran's broader political "machine."[15]

McCarran's wide-ranging connections to the grazing community in Nevada made him the obvious spokesman to address the labor shortage of sheepherders arising during World War II. Nevada woolgrowers saw the solution to their labor problem as the recruitment of Basques from the Old Country, however difficult that might be considering longstanding American immigration restrictions. The employment of Mexican sheepherders was seen as an emergency measure by ranchers until more Basque labor could be obtained. Nevertheless, Mexican labor did not meet the expectations of the stock operators in sheep grazing.

They argued that Mexican labor was not qualified to tend sheep, but the strong subtext suggested discrimination against Mexicans because of their racial background. The preference was for Basques because they were "whiter" than Mexicans (with all the cultural associations that implied), and in general they were considered reliable and honest laborers with a dedication to the care of sheep—the chief capital investment of the industry. Later on, in 1943, John Dangberg, one of the American ranchers advocating the importation of Basques and representing the woolgrowers of the western Nevada, said: "We, as a group, are not prejudiced against any race or creed but you can see that we do not expect much help from the Mexicans." But their desired labor force required higher wages and costs of recruitment than the more close-at-hand Mexican labor. Ranchers, however, seemed willing to bear these costs.[16]

Senator McCarran saw an opportunity to assist the sheep industry in its search for Basque laborers. In the process, John Dangberg became one of the strongest voices in favor of Basque labor in the sheep industry in Nevada and elsewhere in the entire West. Both he and McCarran helped to raise concern over the shortage of Basque labor in the sheep industry to the national level.[17]

"Come Back to the Flocks": The Demands of War

Japan's attack on Pearl Harbor in December 1941 forced Pat McCarran to pay attention to international events, turning his attention away from public lands in the West. Furthermore, the wartime economy, which demanded more livestock production, also forced his attention to the immediate needs of the ranch industry. This included the shortage of agricultural workers—especially sheepherders. As a result, he started looking for ways to open doors to more Basque immigrant workers.

John Dangberg's goals lined up with McCarran's in the increased importation of Basque labor. On April 17, 1942, Dangberg wrote desperately to McCarran urging more attention to the Basque question in the sheep industry of Nevada:

> We are getting more and more anxious about the sheepherder question here. As you know the sheep industry in this section has for years depended on Basque herders. There is a serious shortage of men right now and everyone is feeling it. As for ourselves we have 8,000 sheep ready to go into the hills and are short two

> herders. We have tried Mexicans from time to time but I don't need to tell you that they are not satisfactory when left with sheep in remote mountain pastures.[18]

Dangberg explained to McCarran how the U.S. Employment Service office (this federal government agency, which conducted job placement, was reorganized into the Department of Labor in 1933, after the passage of the Wagner-Peyser Act) could get sheep workers residing in the United States who, in his own opinion, were not qualified. William J. Kane (inspector in charge of the U.S. Department of Labor, Immigration and Naturalization Service) at the Reno immigration office told Dangberg that bringing Basque immigrants from Spain at that moment was "impossible."[19]

These concerns moved McCarran to find legal ways to bring the Basques to the Great Basin. Indeed, the Nevada senator did not take long to reply. On May 4, 1942, McCarran wrote Dangberg promising to do whatever he could on the Basque manpower issue.[20] In the 1930s, Pat McCarran had already tried to find ways to allow groups of Basque immigrants to enter the country. But he had little success. Responding to Dangberg's letter, the Nevada senator went on to request immediate action by the federal government. This correspondence about the sheepherders question marked the beginning of the long cooperation and correspondence between Dangberg and McCarran in an intensive effort to sponsor Basque immigration to the western livestock industry.[21]

Back in Nevada, John Dangberg welcomed McCarran's letter. On May 21, 1942, the Nevada rancher wrote to the senator thanking him for all his support and asserting that the Basque labor matter needed to be solved by the U.S. Employment Service. He repeatedly mentioned how the sheepmen "always hire through the Basque hotels and bars." Despite the difficult situation, Dangberg wrote some optimistic words about the future: "We are getting together on this here now and hope to make our need known to the office in question sometime soon. I will advise you of any progress made."[22]

Six days later, McCarran wrote Dangberg to further declare his support for the recruitment of Basque immigrants: "I am going to do everything in my power to assist with regard to the sheepherder question and in trying to secure Basque herders." Still, McCarran was honest with Dangberg and warned him that there was little possibility of getting Basques from Spain.

Nonetheless, McCarran's letter did not discourage Dangberg as he persevered in finding ways to bring Basques from the Old World.[23]

In the summer of 1942, McCarran went ahead in obtaining more legal information to secure Basque workers in Nevada's sheepherding. On June 1, McCarran sent a letter to John J. Corson, director of the U.S. Employment Service, requesting information to "secure qualified men for sheepherding in Nevada." Then, on June 10, Corson notified the Senator that foreign labor such as Basques depended upon both the Employment Service and the Immigration and Naturalization Service. Thus, all ranchers who wanted to employ Basque herders had to make their requests at the local offices of the Employment Services, and these demands would later be sent to the Immigration and Naturalization Service. Once these applications were made indicating the necessity of importing foreign labor irreplaceable by other workers already in the United States, then the U.S. Employment Service would need to certify that the demand for sheepherders exceeded the supply and that such demand could not be met otherwise. In addition to Dangberg, other ranchers were also having difficulties finding and employing Basque sheepherders. Dangberg began to contact and urge other sheepmen to fill out claims in the local Employment Service offices located in the following Nevada communities: Elko, Ely, Fallon, Las Vegas, Lovelock, Pioche, Tonopah, Winnemucca, and Reno.[24]

The shortage of Basque labor in Nevada's sheep grazing reached its zenith in late 1942. Dangberg was increasingly impatient for action on the sheepherder question. Therefore, he continuously knocked on every possible door and talked to many influential people asking for help to bring Basque immigrants. On December 17, 1942, Dangberg wrote to Ed A. Settlemeyer, another sheepman and an old friend who was chairman of the U.S. Department of Agriculture War Board for the state of Nevada. Dangberg informed Settlemeyer of the critical situation of the Basque sheepherders and asked him to help on this issue: "I believe you would be in a better position to make a constructive suggestion than any man I know of."[25]

Dangberg enclosed a copy of a report showing some information about the labor shortage for the period between 1941 and 1942 for his company. This report was intended to demonstrate the inefficiency of the Employment Service's plan to employ Americans first. At that time, the H. F. Dangberg Land & Live Stock Company operated in the counties of Douglas, Lyon, and Ormsby, and also in Alpine County, California.

The annual production was estimated as follows: 1) number of ewes, 10,000 head; 2) market lambs, 7,500 head; 3) wool crop, 93,000 pounds; and, 4) purebred range bucks, 350 head. Then, the report suggested a difficult situation in the labor pool. It showed both that the number of available sheep workers had significantly decreased and second, those sheepherders already working for the Dangberg ranch were getting older. To Dangberg, the problem was not simply the legal restrictions against importing Basque immigrants; the labor shortage issue spoke to a larger economic problem. With the greater economic diversification and particularly with the defense industries expanding, other new occupations had become more attractive even for the former Basque sheepherders. As Dangberg himself described, there were many Basques who had quit sheep grazing to take higher-paying jobs.

Dangberg also noted some few cases where non-Basque sheepherders who were sent to his ranch from the Employment Service lasted only a few days because of the hard work. For example, R. C. McFarlane of American nationality lasted only two and a half days. On the other hand, a Mexican worker, Frederico T. Mestas, lasted one month and apparently left the job because of the "cold weather." According to Dangberg, moreover, there were workers sent by the Employment of Service who were not qualified for sheep grazing. He noted the following experiences: "One man came in August and had to be returned to the agency because he could not pack a burro. Another man applied Dec. 16 but could not qualify for the same reason." Dangberg noted that the only way to handle this problem was to import Basque labor. He concluded as follows: "Basques are the only qualified men for range herding—they are customarily hired through local Basque hotels and not through the U.S. Employment Service. The draft boards have not taken herders. . . . Herders have left the sheep business for defense jobs in considerable numbers."[26]

On the same day, on December 17, John Dangberg sent a similar letter to General J. White, the state director of Selective Service with office in Carson City, informing of his opinion on the labor shortage in the sheep industry. He wrote that "The situation is now critical and sooner or later will compel the sale of breeding stock to the butcher unless something can be done to persuade the Basque boys to come back to the flocks."[27] Dangberg noted that the day before his daughter had spoken to Nevada governor Edward P. Carville about this problem, and he told her that he was going to discuss it with White.

In the same day, also, Dangberg wrote W. P. Wing, secretary of the California Wool Growers Association, explaining all his concerns in the business and asking for help: "It sounds as though you are both playing Santa Claus to anyone trying to get herders over here in Nevada where the weather is getting colder and colder every day now. However, I suppose we must expect miracles in California sunshine." Dangberg asked Wing if they had any problem in California and where he thought, "the Basques have gone to who were herding for us last year and the year before." The future, in Dangberg's own words, "does not look too good for us." He claimed that new solutions needed to be found: "The U.S. Employment Service has been doing all it can for us and that with unfailing courtesy and patience but the results, as you see, are negligible. The chances of getting herders from Spain are pretty poor although Senator McCarran has said he would go to the bat for us." To Dangberg, however, a main problem was the Basques' lost interest in sheep grazing: "There is only one other 'out' as I see it and that is to persuade the Basques who have gone into defense industries to return to the sheep. However, you may have some better idea." This problem was general and worsening. Dangberg suggested to "persuade" Basque workers regardless of the cost. Because of the lack of Basque labor, as he pointed out, some ranch operators did most of the work themselves, fearing they would have to sell their flocks if no one could herd them. The situation was critical, he added: "Our experience is not unique—all the other range operators are having about the same difficulties—some operators (Basques) are doing their own herding."[28]

By late 1942, Dangberg had put together a strong organization of sheepmen in Nevada. On December 26, from Reno, Ed A. Settlemeyer sent a letter to Dangberg in which he informed him about the difficulties in recruiting Basques from the Old Country. He suggested that the "possibility of getting immigrants from Spain is very distant," and he instead contemplated as a possible alternative the employment of experienced workers from Mexico and bordering states:

> Your suggestions will help a great deal to bring out in the open the facts as they really exist.
>
> Nearly every outfit in our range states is confronted with the same condition as you write about. Food production will be greatly handicapped at best but increasingly so unless reliable help can be found.

> The possibility of getting immigrants from Spain is very distant. By calling the urgency to the attention of the U.S. Employment Service, then, through the Man Power Commission, surplus men with range experience might be moved from Mexico or the border states. I will take this up with the employment service and our board at an early special called meeting.
>
> We are glad to get your suggestions and wish more producers would make the effort you have made, so the facts can be put before those who have been given the power to act.[29]

On December 29, 1942, at 2:00 PM, different sheep operators of western Nevada and some contiguous areas of California met at the Douglas County courthouse in Minden, in order to discuss the shortage of Basque sheepherders. Those woolgrowers came largely from Bridgeport, in California, Smith Valley, Mason Valley, Carson Valley, Truckee Meadows, and the area of Fallon, in Nevada. They primarily grazed their sheep in Nevada Grazing District Number 3 and in the Mono National Forest.[30] In the meeting, attendees reached a unanimous verdict concluding that the annual production was largely dependent upon the Basque immigrant labor.[31] Unless the government allowed the recruitment of Basque immigrant workers, in Dangberg's own words, "range sheep owners will be forced to cut their flocks drastically or sell out thus cutting production of meat and wool." All of them did agree that "Basques are the only men uniformly qualified for this work." Looking ahead, those sheepmen anticipated a "critical shortage" which could develop in the next year lambing season, by April 1943.[32]

At this same meeting in December 1942, the sheepmen founded the Nevada Range Sheep Owners Association (NRSOA). This platform was organized "for the purpose of gathering data relative to the situation and presenting them to the Nevada Congressional delegation and to all other government and private agencies interested in maintaining production." The woolgrowers also intended to use the association to monitor the supply and demand of Basque sheepherders. They formed a political advocacy group through which they aimed to find solutions to the shortage of Basque workers. John Dangberg himself was appointed chairman and D. W. Park secretary of the new association. Additionally, Wilbur Stodieck, member of the U.S. Department of Agriculture War Board and district extension agent for Douglas County Farm Bureau, offered services and facilities of his office to help solve the labor shortage. Stodieck developed a questionnaire that was sent to all sheep owners of western Nevada and some adjacent areas in California. Those ranchers who experienced difficulties with sheep labor or anticipated shortages later in the season were asked to respond to the questionnaire. By filling out and signing this questionnaire, sheepmen immediately became members of the NRSOA. The purpose of this questionnaire was to collect detailed information about the sheep labor in this part of the Great Basin and at the same time demonstrate to the government the labor shortage. Thirty-five percent of the total number of the early NRSOA membership was Basque American.[33]

Dangberg acquired a prominent status within the industry when he became president of the NRSOA. From this moment on, he became the person to contact in order to find possible Basque sheepherders and to find long-term solutions to the problem at state and national level. Dangberg's relations with Senator McCarran became crucial in the effort to increase Basque immigration.[34]

The urgency of the problem of the shortage of Basque immigrant laborers was not just an issue in Nevada. The sheepherder shortage in other western states was becoming distressing as well. Some days earlier, on December 17, Dangberg contacted Dan J. O'Loughlin, secretary of the National Wool Marketing Corporation with their office in Salt Lake City, requesting further information from him about the sheep labor situation in the east-central part of the Great Basin and west of the High Plains. On December 26, O'Loughlin got in touch with several woolgrower associations to get acquainted with the labor shortage in different states. He wrote to various secretaries of woolgrower associations in the Interior West, informing them about Dangberg's report: J. Byron Wilson, secretary of the Wyoming Wool Growers Association; Paul Etchepare, secretary and treasurer of the Montana Wool Growers Association; Mel C. Claar, secretary of the Idaho Wool Growers Association; A. E. Lawson, secretary of the Washington Wool Growers Association; and Albert L. Linger, secretary of the Colorado Wool Growers Association.[35]

Also, on December 31, woolgrower association representatives reported back to O'Loughlin the situation in each state. From McKinley, Wyoming, J. B. Wilson affirmed the criticalness of the labor shortage in the American West and also pointed to a possible solution to this problem by recruiting Basques from Mexico who were exiled and settled there because of the Spanish Civil War: "When I was in Washington

Table 5.1. Early members of the NRSOA

Arambel and Etcheverry	St. John Laborde
Jean Aldax	D. W. Park
Baker and Calvin	H. F. Dangberg Land & Live Stock Co.
John Belaustegui	Chas. Aldabe
W. L. Blackwell	V. S. Connell
Bruce H. Chichester	Gregorio Mendiburu
James Compston	J. B. Saldubehere
J. J. Fleming Co.	Meadow Canyon Sheep Company
C. G. Hussman	Martin Sorensen
Stoddard Jacobsen	Mono Land & Livestock Co. (Joe Sario)
George C. Koenig	Ed. J. Powell Est.
Linscott Bros.	A. L. Dromiack
Domingo Recatune	C. D. & W. Y. Campbell
Mrs. Ida Ruhenstroth	Frank Trosi
Guy M. Terry	Helen Goni
John Uhalde	Holland Livestock Co.
H. W. Whitaker	Alfonso Sario
W. T. Jenkins Co.	H. E. Carter
Warren Simpson	Frank Azcarate
Landa Bros	Antone Saldubehere
L. L. Wedertz	Thomas Ormachea
Warren and Gordon Wedertz	

Source: List of members of the NRSOA, January 16, 1944, Box BSQAP 0115, Nevada Range Sheep Owners Association, Jon Bilbao Basque Library, University of Nevada, Reno, Nevada.

last, I called upon the State Department with the idea that some arrangement might be made whereby we could import Basques from Mexico" However, Wilson said: "I thought that they could probably arrange with stockmen to put up a bond and send an experienced man to Mexico to recruit labor, but the State Department advised me this could not be done because they had made arrangements with the Mexican Government to have the Department of Agriculture recruit labor from Mexico."[36]

According to Wilson, sheepmen could keep the ranch hands they had. However, getting new sheepherders was another issue. He then suggested inquiries with the Employment Service. He pointed out: "We are suggesting to the Department of Agriculture that we are going to need a lot of men from the sheep sections of Old Mexico and perhaps, all together, we may be able to get by although it is going to be pretty serious." Furthermore, he thought that every state in the American West was going to be in the same situation, including Texas. "I was surprised when I was in Texas recently to find they were having a good deal of difficulty in getting men. I had assumed that living as close to the border as they do they would be able to get plenty of men but apparently, they are unable to get men to come over any more and they are suffering from a real labor shortage just as we are." He adopted a cautious position in dealing with the sheepherder shortage. An increase in the wages might be a positive to attract former herders: "I believe average wages in Wyoming for sheep herders will be $100 a month, at least, by spring." In one way or another, Wilson hoped to reach some solution with the Employment Service.[37]

On that same day, Paul Etchepare, a Basque sheepman, on behalf of the Montana Wool Growers Association, sent a letter to Dan O'Loughlin in which he reaffirmed the necessity of importing Basque sheepherders. He further explained how Montana experienced the same situation concerning the sheepherding labor: "Defense industries and attractive wages have taken the greater percentage of our ranch help."[38] Etchepare appeared interested in reading Dangberg's report about the Basque sheepherders problem and was particularly enthusiastic about the point that these immigrants were the only qualified men for range herding.[39]

From Boise, Mel C. Claar, the representative of the Idaho Wool Growers Association, sent another letter explaining the situation of the Basque labor shortage in Idaho. Clearly, Claar noted: "This sheep herder business is serious." Furthermore, he explained how they worked on their own to find solutions since the month of March and alternatively employed Mexican workers to tend sheep. But Mexican labor failed to meet ranchers' expectations: "We are still corresponding and trying to make contacts that might produce some Mexicans, but Mexicans out of Old Mexico are not very satisfactory. The matter is so serious that we are just like a drowning man grasping at a straw." Claar stated also that the labor shortage was significantly constraining sheep grazing in Idaho: "It is our conservative opinion, based on actual knowledge of sales that there are fifty thousand less sheep in Idaho because of owners quitting and another one hundred thousand breeding ewes less on those who have cut down and all because of the labor situation. None of these individuals really want to quit but the unsatisfactory labor, just as reported out of Nevada, is applying here."

Basques had clearly become the more desirable sheep workers over other populations, especially Mexicans. Also, he identified similar problems recruiting Basque refugees living in Mexico, as he spoke on behalf of many ranchers that they were willing to do anything they could to solve the labor issue: "though we lined up a bunch of Spanish-Basque willing to come to the United States, immigration restrictions and the Mexican government has made their importation impossible. We have been working on a theory that if we can ever get the muddle of getting them recruited in Mexico straightened out we will be glad to send a qualified individual who can speak Mexican and Basque, at our own expense, and try to get some employees that would be worth the cost."

The year ended worse for the sheep ranchers than when it started. By the end of 1942, moreover, the frustration among western sheepmen was palpable. Mel C. Claar said that many sheep ranchers from Idaho had not enough men to work in their operations.[40] This led to frustrations with the Employment Service all over the American West. Ranchers prepared to organize their interests to confront the problem. As John Dangberg noted in a letter on December 31 to W. P. Wing, secretary of the California Wool Growers Association, the Nevada ranchmen were taking "further action in the sheepherder situation preparatory to presenting facts and recommendations to our congressional delegation."[41]

Short-Term Solutions

On January 1, not taking a break for the New Year's holiday, John Dangberg wrote a letter to Darrell M. Traugh, forest supervisor of the Mono National Forest with offices in Reno, Nevada. In this letter, he responded to Traugh's previous letter in which he considered the permitted livestock and the issue of commensurability in the government protected lands in Connell's Fish Valley range in the Mono National Forest. Considering the upcoming year, Dangberg also broached the Basque labor problem: "Everything now within the last month or so seems to be getting on a little sounder footing so far as the recognition for agricultural labor needs is concerned, which may have a tendency to help solve the 1943 problem." According to Dangberg, the government's recognition of the sheepherders' crisis came late, after the labor shortage became an acute problem in every agricultural sector. Dangberg claimed that sheepherders were urgently needed from the Old World: "I would like to see some boys brought in from Spain as this is vital to the industry not only for the immediate problem, but has been one of importance now for several years, and unless something can be worked out, it is going to result in a vast change in the whole range sheep business."[42]

Next day, on January 2, after having received the detailed information about the labor situation in other western states, Dangberg also wrote to Dan O'Loughlin his concluding argument:

> The two important facts which are emphasized every time we take a step out here is that the range sheep industry is founded absolutely on the use of Basque herders and that there is right now a critical labor shortage which is going to develop into an avalanche of sales of breeding stock unless we get relief in the next few months. It also looks to us though Spanish Basques would be the answer, whether from Spain or Mexico. The Basques here are of opinion that boys in Spain would come if the immigration law can be changed to admit the necessary quotas. Also for your information our Senator [Patrick McCarran] has been working on something along this line for some time.[43]

As is frequently revealed in his correspondence from this time, Dangberg highlighted Basque immigrants' alleged natural aptitudes for sheepherding. He justified the urgent necessity of recruiting Basque immigrants, by referencing some bad results reported from other operators from the sheep work of other ethnic populations, such as the Irish or Bulgarian workers.[44] Dangberg aimed to demonstrate the scope of the labor shortage and how ranchers could only depend upon Basque sheepherders. Pressures from the wartime economy for increased food production meant that sheep owners in every western state were struggling to find cheap and dependable labor. On January 3, 1943, W. P. Wing from San Francisco wrote O'Loughlin stating that "the labor situation generally in California as it pertains to sheep outfits is acute." As they were increasingly alarmed at the labor shortage situation, they offered higher wages.[45]

Another tactic was to try to lure back Basque sheepherders who had taken jobs in other industries, particularly defense. Some days after its establishment in 1942, the NRSOA had consulted with the local board in Douglas County, Nevada, asking for detailed information about the sheep labor situation—the number of sheepherders registered, whether or not they were still employed in the sheep industry, and what ages they were. The NRSOA discovered that no sheepherder was drafted and that fully two-thirds of them were over forty-five years of age. The information regarding their present occupation—that is, whether or not they were all still in the industry and herding sheep—could not be obtained in the office. The NRSOA suggested identifying those sheepherders who transferred to other employment in the defense industries. However, reaching out to those workers who had taken advantage of the defense industry's job opportunities brought with it several potential obstacles: First, what should they do if those Basques working for other sectors were not willing to return to the sheep industry? Second, was it wise to put unwilling workers out alone for several weeks tending about 1,500 to 2,000 head of sheep? Third, what should they do if the defense industry resists letting those workers go?

Because of these potential obstacles, on January 4, Dangberg noted that "the only 'out' is bringing men from Spain." In order to achieve such goal, the NRSOA was taking every step possible and working closely with their representatives in Congress. In the next few months, Dangberg went to Washington and stayed until the middle of April. He requested that General J. White help Nevada woolgrowers by facilitating the provision of any information from the Selective Service regarding the number—and ages—of any Basque sheepherders that had quit the industry for other employment. He further said that Basques were "also generally skilled iron workers and miners besides being sheepherders" and thereof, he asked White to impress "Washington with the necessity for action if meat and wool are not to be even scarcer." Briefly put, the only way to handle the situation was to employ Basques and only Basque immigrants.[46]

From Yakima, on January 6, A. E. Lawson, secretary of the Washington Wool Growers' Association, wrote a letter to Dan J. O'Loughlin, pointing out that they had started to have problems finding Basque laborers about a year ago and had scoured "the country for sheepherders and packers." The shortage, according to Lawson, was "getting worse all the time." Concerning the Washington range sheep industry, Lawson indicated that the defense industries settled in the Pacific Northwest were abducting all the sheepherders and that there were inequalities in wage rates among the agricultural and military occupations: "Being so close to Portland and Seattle, naturally, the defense industries are taking quite a number of the men." Lawson also noted that the labor shortage was driving prices up. "The average wage paid here now is $100.00 per month for both herders and camp tenders."[47]

From California, on the same day, W.P. Wing suggested John Dangberg contact the Standard Employment Agency of Sacramento and let them know about the sheepherder situation in the Great Basin. Wing knew very well that the problem in the labor pool of the sheep industry was affected by other sectors: "I had a long talk with one of the Basques—in fact two—who had left Nevada sheep herding and have gone into the lumber industry. It is just a question of the wages and conditions of employment." Furthermore, Wing recommended getting in touch with two Basque boardinghouse owners who might know of some sheepherders: Benito Apechechea, owner of the Alturas Hotel in Alturas, California; and Valentin Jaurena, owner of a hotel in Cedarville, California.[48]

Following his advice, John Dangberg wrote two days after receiving Wing's letter to Benito Apechechea requesting Basque herders: "I write you to see if they [those Basque sheepherders working on defense jobs] would consider coming back into the sheep business. I need several herders now and if these boys are interested would be glad to put them on." Dangberg explained to Apechechea the working conditions: "We are paying $110.00 per month—the man must have his own bed and know how to pack a burro. Our foreman is W. Mullins who used to live up in that country and we have a Basque outfit as to herders and camp tenders. We also try to supply our camps with good food and plenty of it." On the same day, Dangberg also wrote a letter to the Murray and Ready Employment Agency of San Francisco, asking for Basque sheepherders seeking employment in Nevada at any time. In this letter, Dangberg explicitly said that the workers had to be Basques: "A man coming to us as a herder must be *Basque* [*sic*], must understand how to pack a burro, how to cook for himself and he must have a bed roll." He desperately added that they did not pay "transportation but if a man is a good herder, we will consider advancing fare and then holding it out on his wages." The same day, W.P. Wing wrote to Francis Wood urgently seeking his help to find Basque labor available in the West. He mentioned the cases of two prominent stock companies: the H.F. Dangberg Land and Livestock Company of Minden and the company owned by J.J. Fleming of Wendel, Lassen County, in California. Wing also stated his opinion to Wood that in order to facilitate the sheepherders' recruitment process, he advised the establishment of mechanisms to allow direct communication between the sheepmen and the Employment Service, without the intermediary role of the woolgrower associations.[49]

On January 16, 1943, the NRSOA held a meeting in Minden, in which local sheepmen, as well as representatives of the Employment Service, Forest Service, and the local war boards came together to discuss the latest progress regarding the labor recruiting process. They presented the results of the first analyses of questionnaire data sent on December 29 to the sheep operators of western Nevada. The purpose of the questionnaire was to form the basis of factual data. 87 percent of the owners in western Nevada and adjacent areas in California—representing 145,000 head of breeding ewes—answered and returned this survey to the NRSOA. On the questionnaires, most of the owners showed their dissatisfaction with both the Mexican labor and the sheepherders available through the Employment Service. According to D.W. Park, secretary of the NRSOA, "owners have not found Mexicans satisfactory and have no confidence that they can continue operating if they are forced to accept Mexican herders in this crisis." In the meeting, sheepmen reached a consensus that "in as much as the industry has been built on the skill of the Basque herders, the Basques are the only men who can keep us in production."[50]

To help federal agencies understand the labor needs of the Nevada woolgrowers, they reported to their representatives the difficulties in getting Basque immigrants from Europe, along with other issues affecting labor shortages. Nevada woolgrowers adopted a resolution reporting the sheep labor situation and their challenges. Based on the questionnaires sent to ranchers, the NRSOA estimated that there were 180 Basque sheepherders employed at that time. For the next lambing season, they aimed to employ 330 sheepherders and for the spring range time, 230 ranch hands, between herders and packers. Besides immigration restrictions and better job opportunities in wartime, aging became part of the problem: "Fully 50% are of an average age of 55 years or over and many are past 50, making for serious inefficiency due to unexpected illness etc." The report showed that: first, 3 percent of the "skilled herders" were drafted; second, 8 percent were unable to continue working because they were "too old or too ill;" and third, 13 percent of the Basque labor had quit sheepherding and transferred to defense industries. In such a meeting, those Nevada sheepmen unanimously concluded that:

"Unless 70 additional skilled and reliable range herders are available by May 1, which represents a 30% increase over present numbers employed, at least 30% of the breeding ewes now on hand, or 48,000 head producing annually approximately 3,500,000 pounds of marketable lamb and 480,000 pounds of wool will have to go to the butcher."[51]

All in all, the NRSOA report stressed that sheep production in the western section of Nevada growth rates would be constrained if the Basque labor question was not soon solved. Immediately after the meeting, John Dangberg wrote letters to their representatives in the Senate and Congress—Patrick McCarran, James G. Scrugham, and Maurice J. Sullivan—to press for haste in sponsoring Basque immigration and, in D.W. Park's own words, "to give a true picture of the critical need for trained men to enable us to carry on."[52]

Summary

The development of the sheepherding industry at this time and its labor needs encouraged the increasing articulation of specific perspectives on Basque immigrants, particularly with regard to being innately skilled in sheepherding. By the time the United States entered World War II, Nevada's sheep industry and that of other western states had run out of capacity to meet wartime demands. The Second World War revealed a Basque immigrant labor shortage in the sheep industry of Nevada and the American West and Congress's restrictive immigration measures put in place during the mid-1920s had significantly reduced the amount of Basque labor available for sheep grazing. By the late thirties and early forties, furthermore, most of the earlier Basque immigrant sheepherders had reached old age, retired, or moved on to another, better job. Many Basque sheepherders were drawn to job opportunities for wartime employment all over the West. As a result, employing Basque immigrants for sheepherding became even more difficult as a result, something that prompted a strong response from the industry focused on recruiting specifically Basque sheepherders.

During World War II when demand for all sorts of labor rose, the lack of labor in the sheep industry became increasingly critical, and John Dangberg's organizing efforts and leadership of a group of sheep ranchers of western Nevada were heavily directed at recruiting more Basque immigrants despite the restrictive immigration laws. Much of those efforts were aimed at finding ways to get around those restrictions, something which is clearly revealed in their correspondence to Nevada's senator McCarran, asking for help to secure Basque immigrant labor for sheepherding. As part of these organizing efforts, Nevada western woolgrowers established the Nevada Range Sheep Owners Association (NRSOA) to address their collective interests in finding Basque sheepherders and enlisting the help of their congressional delegation in Washington, D.C., something which underscored the broad urgency within the industry for action to address the challenges ranchers were facing.

During the last part of his life from 1942 to 1954, McCarran would continue to serve his sheep-grazing constituency's efforts to bring Basque immigrant labor to the United States. As McCarran gained power and influence in the Senate, he would seek ways to open the immigration doors to Basques, something that I explore in depth in chapter 6. The scarcity of Basque immigrant labor began to be recognized as a serious handicap to the development of the sheep industry in Nevada and the West as soon as McCarran took the matter to the Capitol.

As the war continued, sheep ranchers from Nevada and other western states also continued struggling to find Basque sheepherders to work in their operations. However, their extensive organizing and advocacy efforts would eventually pay off. The next chapter examines how these efforts unfolded in the final years of the war, and how it was that the Western woolgrowers, with the help of Senator McCarran, were able to bring additional Basque immigrants from Spain to work in the sheepherding industry.

Notes

1. Bill Johnston, "In Defense of Sheepherders," *Idaho Wool Growers Bulletin* 20, no. 12 (June 5, 1940): 4.
2. Ibid.
3. Ranchers and others involved in the recruiting process of the Basque immigrant labor commonly used the term "import" to refer to the admission of this type of labor to the United States. The usage implies commoditization of these immigrant sheepherder workers. This study also employs this term at times to refer to the process of bringing immigrant workers to the United States, Patrick McCarran to John Dangberg, May 4, 1942, Collection no. 13, Box 38, Dangberg Correspondence 1942, McCarran Papers.
4. Jerome E. Edwards, "Nevada Power Broker: Pat McCarran and His Political Machine," *Nevada Historical Society Quarterly* 27, no. 3 (Fall 1984): 182–98; *Labor*, August 23, 1938, 1.

5. Gibson "Gib" Crockett, "Senator Pat McCarran of Nevada," *Sunday Star*, June 22, 1941.
6. Alfred Steinberg, "McCarran: Lone Wolf of the Senate," *Harper's Magazine* (November 1950): 89–95; Gilman M. Ostrander, *Nevada: The Great Rotten Borough, 1859–1964* (New York: Alfred A. Knopf, 1966), 152–56.
7. George Creel, "Under the Underdog," *Collier's*, April 6, 1935, 26; Edwards, *Pat McCarran*, 94–121; Peffer, *Closing of the Public Domain*, 247–78.
8. Gerald D. Nash, *The American West Transformed: The Impact of the Second World War* (Lincoln: University of Nebraska Press, 1985), 23, 85.
9. "You Do the Fighting and We'll Produce the Lambs and the Wool," *Idaho Wool Growers Bulletin* 21, no. 16 (August 12, 1942): 13.
10. Elliott, *History of Nevada*, 312–18.
11. *Idaho Wool Growers Bulletin* 21, no. 26 (December 17, 1941); *Idaho Wool Growers Bulletin* 22, no. 2 (January 28, 1942); *Idaho Wool Growers Bulletin* 22, no. 4 (February 25, 1942); *Idaho Wool Growers Bulletin* 21, no. 22 (November 4, 1942); *Idaho Wool Growers Bulletin* 21, no. 25 (December 16, 1942); *Idaho Wool Growers Bulletin* 21, no. 26 (December 30, 1942).
12. Patrick McCarran to John Dangberg, January 19, 1944; John Dangberg to Patrick McCarran, January 14, 1944, Collection no. 13, Box 39, Dangberg Correspondence 1944, McCarran Papers.
13. Douglass and Bilbao, *Amerikanuak*, 303–06.
14. Merrill, *Public Lands and Political Meaning*, 169–92; R. McGreggor Cawley, *Federal Land, Western Anger: The Sagebrush Rebellion and Environmental Politics* (Lawrence: University Press of Kansas, 1993), 73–74, 93, 111.
15. Paul Laxalt, *Nevada's Paul Laxalt: A Memoir* (Reno, Nev.: Jack Bacon & Co., 2000), 10; Edwards, *Pat McCarran*, 130–31; on "McCarran boys," see also biography of Alan Bible, Gary E. Elliott, *Senator Alan Bible and the Politics of the New West* (Reno: University of Nevada Press, 1994); and Richard O. Davies, ed., *The Maverick Spirit: Building the New Nevada* (Reno: University of Nevada Press, 1999).
16. John Dangberg to Attendance Mr. Moore, February 17, 1943, Collection no. 13, Box 39, Dangberg Correspondence 1943, McCarran Papers; Melvin W. Reder, "The Theory of Occupational Wage Differentials," *American Economic Review* 45, no. 5 (December 1955): 834–35; by that time, American sources continued to see Basques as a distinctive racial type, an interpretation that endured for years. In 1942, Saxton Bradford, a reputed newspaper editor who would later serve as the counselor at the U.S. Embassy in Madrid, Spain, from 1955 until 1957, wrote an article on his adventures and views after living in a Basque immigrant community in the Interior West for several days. Bradford contended that Basques were racially qualified more than others to work in the West with sheep. Bradford asserted: "Before the Basques came it was hard to find men who could be trusted to take out a band of sheep and bring it back without loss." Sheep ranchers, he pointed out, "quickly recognized the difference between these tall, capable, serious Spanish-speaking youngsters and the carefree Mexican-Indians and renegade

whites who rode and herded for them." Saxton E. Bradford, "Sons of the Pyrenees in the Northwest," *Travel* 78 (September 1942): 14.

17. John B. Dangberg was one of the sons of the prominent ranching family in western Nevada. He was born to Heinrich Friedrich and Margaret Gale Dangberg in 1871. In 1855, H. F. Dangberg Sr. settled and established a stock ranch business in Carson Valley. During the 1870s, the family operation grew and prospered, accumulating ranch lands and water rights. In 1904, H. F. Dangberg Sr. passed away. After his death, everything remained in the hands of his family and their sons ran the family business. At that time, the H. F. Dangberg Land & Live Stock Company comprised 33,473 acres. From 1904 to 1953, John B. Dangberg was the president of the H. F. Dangberg Land and Live Stock Company and he administered the business throughout the first half of the twentieth century until his death in 1958. He also managed the sheep operation following his father's path. From the late nineteenth century, the Dangbergs had been employing Basque immigrants to work tending their sheep flocks. John Dangberg judged Basques as the only qualified men for sheep grazing. Perhaps, Dangberg's priority of employing Basque labor in his livestock corporation was best reflected in March 1940 when Arthur Rothstein, as part of the Farm Security Administration (FSA) photography group, portrayed the hard reality of various Basque sheepherders on the Dangberg Ranch in Douglas County, Nevada. Steve Achard and Conrad Buedel, *Lost Legacy of Carson Valley: The Rise and Fall of the H. F. Dangberg Ranching Empire* (Minden, Nev.: Dangberg Partners, 2011), 12–22, 40–105; Clifford C. Walton, *Capitol's Who's Who for Nevada, 1949–1950* (Portland, Ore.: Capitol Publishing Company, 1949), 278; Report of H. F. Dangberg Land & Live Stock Company about the labor situation in the Company's range grazing between 1941 and 1942, Collection no. 13, Box 38, Dangberg Correspondence 1942, McCarran Papers; Arthur Rothstein Photographic Collection, March 1940, Farm Security Administration, Office of War Information Photograph Collection, Library of Congress, Prints and Photographs Division, Washington, D.C.
18. John Dangberg to Patrick McCarran, April 17, 1942, Collection no. 13, Box 38, Dangberg Correspondence 1942, McCarran Papers.
19. Ibid.
20. Patrick McCarran to John Dangberg, May 4, 1942, Collection no. 13, Box 38, Dangberg Correspondence 1942, McCarran Papers.
21. The problem seemed to be the same in other western states. From Boise, on May 20, 1942, the Idaho Wool Growers Association published in its newsletter that "in many communities there has been formed farm labor subcommittees through which by cooperation those in need of various kinds of agricultural labor are attempting to work out some solution." Thus, responding to the sheepherder shortage, many Idaho sheep ranchers organized themselves through county subcommittees by lining up community crews to get the work done. "Labor Problems," *Idaho Wool Growers Bulletin* 22, no. 10 (May 20, 1942): 6.
22. In this same letter, Dangberg further noted to McCarran the importance of the work in the issue pertaining to the reclamation site withdrawals from the public

domain, John Dangberg to Patrick McCarran, May 21, 1942, Collection no. 13, Box 38, Dangberg Correspondence 1942, McCarran Papers.

23. Patrick McCarran to John Dangberg, May 27, 1942, Collection no. 13, Box 38, Dangberg Correspondence 1942, McCarran Papers.
24. John Corson to Patrick McCarran, June 10, 1942; Patrick McCarran to John Dangberg, June 16, 1942; List of Addresses of United States Employment Service for Nevada, June 10, 1942, Collection no. 13, Box 38, Dangberg Correspondence 1942, McCarran Papers.
25. John Dangberg to Ed Settlemeyer, December 17, 1942, Collection no. 13, Box 38, Dangberg Correspondence 1942, McCarran Papers.
26. Report of the H. F. Dangberg Land & Live Stock Co. December 17, 1942, Collection no. 13, Box 38, Dangberg Correspondence 1942, McCarran Papers.
27. John Dangberg to General J. White, December 17, 1942, Collection no. 13, Box 38, Dangberg Correspondence 1942, McCarran Papers.
28. John Dangberg to W.P. Wing, December 26, 1942, Collection no. 13, Box 38, Dangberg Correspondence 1942, McCarran Papers.
29. Ed Settlemeyer to John Dangberg, December 17, 1942, Collection no. 13, Box 38, Dangberg Correspondence 1942, McCarran Papers.
30. An area under administration of the Grazing Service, Nevada Grazing District No. 3, and the Mono national forest. An affidavit submitted by John Dangberg and D.W. Park to the Immigration and Naturalization Service, June 30, 1943, Collection no. 13, Box 39, Dangberg Correspondence 1943, McCarran Papers.
31. In 1943, the NRSOA members ran 165,000 head of breeding ewes in the Nevada Grazing District Number 3 and Mono national forest. These ranchers produced annually over 10,000,000 marketable lambs, 1,568,650 pounds of wool, and 110,000 pelts. Resolution adopted by the NRSOA on January 16, 1943, Collection no. 13, Box 39, Dangberg Correspondence 1943, McCarran Papers.
32. Foundational letter of the Nevada Range Sheep Owners Association (NRSOA), December 30, 1942, Collection no. 13, Box 38, Dangberg Correspondence 1942, McCarran Papers.
33. Ibid.
34. Foundational letter of the Nevada Range Sheep Owners Association (NRSOA), December 30, 1942, Collection no. 13, Box 38, Dangberg Correspondence 1942, McCarran Papers; John Dangberg to General White, January 4, 1943; W.P. Wing to John Dangberg, January 6, 1943, Collection no. 13, Box 39, Dangberg Correspondence 1943, McCarran Papers.
35. Dan J. O'Loughlin to J.B. Wilson, December 26, 1942; Dan J. O'Loughlin to John Dangberg, December 31, 1942; J.B. Wilson to Dan J. O'Loughlin, December 31, 1942; Paul Etchepare to Dan J. O'Loughlin, December 31, 1942; M.C. Claar to Dan J. O'Loughlin, December 31, 1942, Collection no. 13, Box 38, Dangberg Correspondence 1942, McCarran Papers; A.E. Lawson to Dan J. O'Loughlin, January 6, 1943; Albert L. Linger to Dan J. O'Loughlin, January 7, 1943, Collection no. 13, Box 39, Dangberg Correspondence 1943, McCarran Papers.

36. J.B. Wilson to Dan J. O'Loughlin, December 31, 1942, Collection no. 13, Box 38, Dangberg Correspondence 1942, McCarran Papers.
37. Ibid.
38. Paul Etchepare did not know that John Dangberg had written the letter that had been sent to Dan O'Loughlin and that the latter had forwarded to him and other woolgrower secretaries with Dangberg's name omitted, Paul Etchepare to Dan J. O'Loughlin, December 31, 1942, Collection no. 13, Box 38, Dangberg Correspondence 1942, McCarran Papers.
39. Ibid.
40. M.C. Claar to Dan J. O'Loughlin, December 31, 1942, Collection no. 13, Box 38, Dangberg Correspondence 1942, McCarran Papers.
41. J.B. Wilson to Dan J. O'Loughlin, December 31, 1942; Paul Etchepare to Dan J. O'Loughlin, December 31, 1942; M.C. Claar to Dan J. O'Loughlin, December 31, 1942, Collection no. 13, Box 38, Dangberg Correspondence 1942, McCarran Papers.
42. John Dangberg to Darrell M. Traugh, January 1, 1943, Collection no. 13, Box 39, Dangberg Correspondence 1943, McCarran Papers.
43. John Dangberg to Dan J. O'Loughlin, January 2, 1943, Collection no. 13, Box 39, Dangberg Correspondence 1943, McCarran Papers.
44. Ibid.
45. W.P. Wing to Dan J. O'Loughlin, January 3, 1943, Collection no. 13, Box 39, Dangberg Correspondence 1943, McCarran Papers.
46. John Dangberg to General J. White, January 4, 1943, Collection no. 13, Box 39, Dangberg Correspondence 1943, McCarran Papers.
47. A.E. Lawson to Dan J. O'Loughlin, January 6, 1943, Collection no. 13, Box 39, Dangberg Correspondence 1943, McCarran Papers.
48. W.P. Wing to John Dangberg, January 6, 1943, Collection no. 13, Box 39, Dangberg Correspondence 1943, McCarran Papers.
49. John Dangberg to Benito Apechechea, January 8, 1943; John Dangberg to Murray and Ready Employment Agency, January 8, 1943; W.P. Wing to Francis H. Wood, January 8, 1943, Collection no. 13, Box 39, Dangberg Correspondence 1943, McCarran Papers.
50. D.W. Park to Patrick McCarran, January 16, 1943; Resolution adopted by the NRSOA on January 16, 1943, Collection no. 13, Box 39, Dangberg Correspondence 1943, McCarran Papers.
51. Resolution adopted by the NRSOA on January 16, 1943, Collection no. 13, Box 39, Dangberg Correspondence 1943, McCarran Papers.
52. D.W. Park to Patrick McCarran, January 16, 1943; D.W. Park to James G. Scrugham, January 16, 1943, Collection no. 13, Box 39, Dangberg Correspondence 1943, McCarran Papers.

6

The Indispensable Basque Sheepherder

Senator Patrick McCarran and the Sheep Lobby, the Exclusion of Mexicans, and the Recruitment of Basque Immigrants in the Western Sheep Industry during World War II

By early 1943, the Nevada and western sheep industry continued to be haunted by the issue of Basque labor shortage. In June, from San Francisco, W. P. Wing wrote to John Dangberg: "We note that apparently Texans are going across the line and hiring Mexicans for sheep herding."[1] This note from Wing reflected how ranchers in the American West considered Mexicans—the available labor force—as deficient and undesirable sheepherders. Discrimination against Mexicans was widespread and persisted. While woolgrowers discriminated against Mexicans as undesirable labor, at the same time they enthusiastically continued glorifying and praising Basque immigrants as good sheepherders. Yet, sheep ranchers exaggerated the racial otherness of Mexicans that made Basque immigrants much more desirable subjects.

During those years, then, Nevada and other western ranchers deliberately created racial and cultural divisions between Basques and Mexicans to justify the necessity of opening the doors for Basques. Ranchers' discourse appears to be based on cultural differences, but rooted in constructions of race. The U.S. government officials promoted and encouraged the recruitment of Mexicans through the Bracero Program of 1942—an agricultural worker program that was established initially to meet the general wartime labor shortage that had granted rights-to-work to Mexican nationals in the United States. But woolgrowers from Nevada and other western states were not pleased with the government's suggestions to recruit Mexican workers for sheep grazing.

The present chapter explores how Nevada and western woolgrowers continued deeming Basque sheepworkers indispensable to meet wartime production requirements. From 1943 to the end of World War II in 1945, western woolgrowers united to petition government officials to allow the recruitment of further Basque immigrants to work in the sheep industry. With the help of Senator Pat McCarran, Nevada's sheep ranchers headed and coordinated other western woolgrowers' associations to recruit a group of Basque immigrants to work as sheepherders in the American West.

The Claims of Unity

Efforts by western sheepherders to address the Basque labor shortage were ongoing in the midst of the complex international political climate. For instance, right around the same time that the NRSOA were meeting in Minden in the final week of December 1942, President Franklin D. Roosevelt and British Prime Minister Winston Churchill were about to meet in Casablanca, Morocco to discuss their strategy in the fight against the Axis powers, eventually declaring on January 24 their demand for an "unconditional surrender" from the Axis powers. Despite the complexity of the international political and military situation, the Basque labor shortage was still discussed that January in the Seventy-Eighth U.S. Congress. On January 21, Homer Mooney, secretary of Nevada Congressman Maurice J. Sullivan, promised in a letter to John Dangberg that the manpower problem was going to be a live issue in such a session of Congress. He praised Dangberg's previous letter about the Basque labor shortage and further said that they were "informed that Congressmen in California are also deeply interested in the problem."[2]

The War Manpower Commission was for its part studying further ways and means of securing sheepherders in the Great Basin. Nonetheless, their margin of maneuver was very limited because of the restrictions on immigration applied primarily to Spain. On January 22, 1943, Tom Jolly, manager of the War Manpower Commission of Reno, noted to Dangberg the following: "It may be possible that some of this type of labor may be available to us at the army camps situated at Las Vegas and Wendover, and if in your opinion, we would be justified in attempting to secure labor from these two camps, from among those who are now being or will be in the near future, discharged by reason of their occupational classification together with their age range."[3]

On January 28, 1943, after the National Wool Growers Convention in Salt Lake City, W.P. Wing wrote Dangberg, outlining three possible options to address the problem of importing Basque immigrants from Europe: bring Basques already living in Mexico in under the Mexican immigration quota; find a way to allow Basques to emigrate from Spain; or lastly, the importation of experienced Mexican sheepherders.[4] In the communications that occurred around these strategies, however, another issue was raised: the ideological persuasions of potential Basque workers. Basque exiles in Mexico who escaped from Francisco Franco's rule in Spain were generally considered leftists in the United States. Moreover, some questioned whether Basque exiles in Mexico, Latin America, or those willing to resettle in America were communists, and so considered a threat. There were even some who took this further and were opposed to letting Basque refugees into the United States because of these communist fears, fears which were already being stoked in the United States during the wartime years. Just as the tensions between United States and the Soviet Union increased after the end of the war, this issue would become more problematic when sheep ranchers had to justify the necessity of recruiting Basque immigrants.

As the war dragged on, the labor shortage and public-land regulation issues facing woolgrowers in eastern Nevada were still much the same as before. On January 29, 1943, the Eastern Nevada Sheep Growers Association (ENSGA) urged unity among sheepmen in their efforts to remedy these issues. From Ely, Nevada, D.A. Hughes, secretary of the ENSGA, wrote a long letter to John Dangberg, in which he stated the main concerns affecting sheepmen in this part of the state. Hughes, speaking on behalf of the Eastern Nevada sheepmen, demanded more local solutions and joint efforts among sheep ranchers. The ENSGA thought that "by organization of local association, each group, or local association, would be equipped to handle their local problems effectively." Looking at the many crises that the sheep industry faced, they believed that there was "the need of '*UNITY*'," [*sic*] and further, in Hughes' own words, "through organization much can be accomplished for the good of our industry." Furthermore, as Hughes explained, the ENSGA also considered essential the unity among all Nevada woolgrower associations to reach common solutions to the problems affecting the sheep industry. He also saw as problematic the possibility of a future labor organization among sheepherders: "In the near future, the [sheep] industry will be faced with an 'organized labor' problem—[*sic*], I understand efforts are now being made to organize herders in some parts of the West, at this time. What are we going to do about it? . . . Many more serious problems are to arise, which will be 'headaches,' unless the machinery is properly set up." Looking at the whole economic picture, in their opinion, Nevada woolgrowers must cooperate through the establishment of a statewide association to address the Basque labor shortage and other issues. This was the way other woolgrowers from other western states, such as Utah and Idaho, followed and, in Hughes's words, the way Nevadan sheepmen must do it. He cheered: "We can. We should. We must.—Let's do it. [*sic*]"[5]

From Salt Lake City, on January 29, Dan O'Loughlin wrote back a letter to John Dangberg informing him of activities in Utah and eastward states in the High Plains. Earlier in January, at the National Woolgrowers Convention in Salt Lake City, the sheepherder situation was given much attention. After the woolgrower's meeting and a visit in Denver, O'Loughlin confirmed that sheepmen were "having this same trouble everywhere and the whole situation is being eagerly discussed at meetings and in discussions printed in various magazines and wool growers bulletins." O'Loughlin enclosed a copy of a form to request sheep workers submitted by the Wyoming Wool Growers Association and published in the *Wyoming Wool Grower* magazine.[6]

On February 1, 1943, John Dangberg wrote back to W.P. Wing about his latest letter on the different remedies to bring Basques into the United States and the issue of the alleged communist affiliation of the Basque exiles in Mexico. As the next lambing season approached, Dangberg and other Nevada woolgrowers were getting increasingly desperate to find dependable labor. As Dangberg wrote: "Our own situation is getting worse and worse with fewer and fewer qualified men to do the herding. Just now, about two weeks before lambing, we are losing a number of ewes that were injured by being walked many unnecessary miles over the desert by one of these herders—or would-be-herders—that are hiring out through the employment agencies." Among the three remedies suggested by Wing in his earlier letter, to Dangberg, recruiting those Basques living in Mexico was the only temporary remedy to fill their labor needs. Furthermore, Dangberg tried to avoid the political issues mentioned by Wing and came to the Basques' defense:

> If these fellows have herded, want to work and are well treated I doubt if their political affiliations will be very important. As

> someone remarked here, they had to choose between being Fa[s] cists or Communists. As far as this business of communism is concerned, the Mexicans are just as apt to be of that persuasion and they, as far as I have seen them here, carry conduct usually associated with what we call communism right over into their work. Of course, in stating anything about the Basques in Mexico, I do not want to suggest that exploration of the situation with respect to importations from Spain be not considered; my thought is rather that those in Mexico offer the only relief possible for this spring.[7]

During this time, the Employment Service in Nevada had developed a recruiting plan for sheep agricultural workers in Mexico. This plan considered recruiting Mexican nationals to work in the western United States. On February 2, from Reno, Nevada, Tom Jolly of the Employment Service gave Dangberg a telephone call and discussed the labor situation. Jolly explained to Dangberg that for the service of recruiting Mexican workers, the federal government charged $5 per "head" to the sheepman. Next day, Dangberg wrote a letter back to Jolly in which he agreed with the conditions and the fee imposed by the government for employing sheepherders. However, this effort to recruit Mexican labor was not enough for the woolgrowers, and they continued complaining of the Basque labor shortage.[8]

Ten days later, on February 12, Dangberg wrote O'Loughlin a letter in which he related how the Nevada woolgrowers celebrated the work that the latter was doing in the effort to solve the sheepherder question and "in getting everybody going at the same thing at once." On the same day, Dangberg also addressed a letter to D.A. Hughes, secretary of the ENSGA, showing his readiness to find a common compromise and total endorsement of the establishment of an organization plan at the state level: "I trust your organization has stated its position on the herder question to our representatives in Congress and has put the weight of its influence behind them to the end that we get action from the State Department on getting herders in here."[9]

That day as well, Paul V. McNutt, chairman of the War Manpower Commission, informed Pat McCarran "that the question of the importation of Basque herders is being given careful consideration." Since late January, however, the full responsibility for the recruitment, transfer, and placement of agricultural workers had been transferred to the Department of Agriculture, with the purpose of coordinating and centralizing under one office all matters affecting the agricultural development. From this moment on, then, as McNutt pointed out to McCarran, the Department of Agriculture was charged with the responsibility of managing all the issues and difficulties confronting the sheep industry, including Basque immigrants' recruiting problems. Upon the transfer of such duties to the Department of Agriculture, McNutt took the liberty of referring to McCarran's letter about the question of the Basque sheep workers. He felt confident enough that "this question will be given careful consideration by Secretary [Claude R.] Wickard's organization." Nevertheless, McCarran's relationship with the Department of Agriculture and the Department of the Interior had deteriorated during his battle against the Grazing Service and its proposed grazing fee policy.[10]

On February 15, 1943, after he interchanged correspondence with different secretaries of woolgrower associations and talked personally with many sheepmen in Utah, Colorado, Idaho, and Wyoming, Dan O'Loughlin reported to Dangberg his thoughts on the Basque sheepherder question, copying out several quotes from various western woolgrower association secretaries to show the dimension of the problem. He contended: "In as much as the army is taking all the men into their organizations the labor question, of course, will become more serious but there seems to be a feeling all round that it will work out alright, and while I have my own personal opinion on the subject, I hate to be too critical because it might look like an attempt to show I knew more about running a war than the people who are actually doing it."[11]

O'Loughlin criticized the ambiguous position of the federal government on the sheep labor issue. Briefly put, while the government encouraged and demanded production to meet the wartime economy, it did not take extraordinary measures or implement new policies to confront the administrative constraints affecting the sheep industry, such as the importation of Basque immigrant labor.

On the same day, while Senator McCarran was leading investigations against the Grazing Service in his position on the Committee on Public Lands and Surveys in the American West, his secretary, Eva Adams wrote to D.W. Park, secretary of the NRSOA, informing him about the administrative changes in the procedure for the recruitment of immigrant labor as the result of the mentioned transfer from the

Employment Service to the Department of Agriculture. In any event, Adams also said that McCarran wanted to keep him informed of Nevada woolgrowers' problems and particularly, the senator wanted to know how the issue of Basque labor was developing in the Great Basin.[12]

Although the international crisis increased immigration restrictions to the United States, plans to hire Basque herders in Mexico began. The NRSOA had prepared an application to request one hundred workers from Mexico through the Employment Service office of Reno. In this request, the Nevada western woolgrowers made clear that they preferred to employ those Basque exiles living in Mexico rather than Mexican nationals. On February 23, 1943, Dangberg wrote McCarran informing him of this and other issues. Dangberg said that language problems among workers could serve as a strong argument to convince the federal officers to take measures to bring only Basque immigrants. In his claim of an urgent recruitment of Basque sheepherders, speaking on behalf of western Nevada woolgrowers, Dangberg argued that ethnic community cohesion among Basque immigrants improved and made secure the working environment. He said that federal authorities,

> . . . should give some consideration to the language problem which will arise with Mexicans direct from Old Mexico. These men will speak only Spanish and will be hard to tell about range lines and range use for the reason that we have few Spanish-speaking people here. We can handle Basques, whether Spanish or French, because they all use the Basque language. There are a far greater number of French Basques here than Spanish, although Spanish Basques are just as good for the herding job as the French.[13]

Dangberg persevered to argue for the exclusion of Mexicans. His point of view implied that diverse and mixed working environments based on language issues led to disunity among the agricultural labor. He said that employing Mexican labor on a Basque-dominated sheep operation could disrupt the working environment because of communication problems between Mexicans and Basques from the French provinces, whereas those Basque exiles living in Mexico who came largely from Spain could speak both Basque and Spanish languages. Dangberg also argued that Basque laborers showed cohesiveness in their common ethnicity and language, something which made them work as a community and a cooperative and effective labor force—more so, in his opinion, than Mexican workers. By mid-March, as the NRSOA eagerly waited for a response about the sheep owners' application from the Employment Service office in Reno, they still had not heard anything. This greatly frustrated the sheep operators.[14]

By late February, at Dangberg's request, the ENSGA had also taken the matter of the Basque labor shortage up with the Nevada congressional delegation with the view of securing action in Washington, D.C. The cooperation between the NRSOA and the ENSGA began to work toward a common objective: bring Basques to the sheepherding jobs in Nevada. Nevada woolgrowers not only pressured their representatives, but also contacted representatives of California because there were members of the NRSOA who grazed sheep in the adjacent ranges in Mono County. On March 15, Dangberg addressed Harry L. Englebright, California congressman. Dangberg explained in detail how "critical" the situation was. "Every man I know who has sheep in Mono County," he said, "is short of herders." Dangberg requested help from Englebright to support McCarran's efforts to import Basque immigrants: "We hoped that by taking action in January we would have men here by the time the situation became critical but our effort has bogged down somewhere in Washington in the bureaus. I know you know how to get action if anything can be done and I feel sure that the help of the California delegation will do a lot towards making Senator McCarran's efforts effective now." That same day, Dangberg's fellow sheepman D.W. Park wrote to McCarran thanking him for all his help and his comprehension of the urgent "need for Basques." As the lambing season proceeded, Park wrote, however, to say that sheep ranches were still in need of labor for summer herding: "I can assure you that the situation is acute—there are no replacements available for sick men, the present men are, in many cases, old and the ewes and lambs ready to go on spring ranges. We have produced the lambs—now we need the men to bring them to marketable age."[15]

On March 21 and 22, the Nevada Wool Marketing Association held its annual stockholders' meeting in Ely. Many woolgrowers from different corners of Nevada attended that meeting, where they exposed the problems affecting the sheep industry. They drafted a resolution pointing out a 30 percent increase in labor costs. In addition, they reported other increased production costs: Forest grazing costs over 22 percent, concentrated foods 30 to 40 percent, transportation 25 percent, and management 50 percent. Sheepmen blamed the federal government for increasing regulation.[16]

The NRSOA continued to put pressure on the Employment Service to recruit Basque sheepherders. From Washington, on March 30, 1943, at 9:25 AM in the morning, McCarran sent a telegram to Dangberg advising him that he was working continually on the Basque sheepherder question. He confirmed some good results from several conferences with the Labor Division of the Department of Agriculture about the requested one hundred workers from Mexico: "Conferences with Agriculture Department of Labor Division have brought about immediate activity for importation of one hundred sheep herders from Mexico hope to have majority Spanish or French Basques doing everything in my power along this line kind regards." Dangberg was pessimistic about the admission of those one hundred herders from Mexico. Besides the possibility of bringing Basques from Mexico, he further remembered that they could recruit former Basque herders from among military reservists. Dangberg also suggested that they try to engage Basques on ships regularly plying the route between Spain and any South American countries—such as Colombia and Venezuela—and from there, bring them to the United States. This alternative needed, however, a special provision from Congress. With the worsening situation, Dangberg urged McCarran to send a quick response. Dangberg said the situation was so critical that some ranchers were being forced to sell off part of their sheep (in April, for instance, the Edmund Powell Estate had to sell one thousand head of ewes) and many others were confronting problems in maintaining production due to the labor shortage.[17]

In order to request Basque immigrants and release them from the army, ranchers individually were required to make an affidavit that would serve as evidence of the need for labor. By this affidavit, a rancher must specify the following information: number of sheep grazed, operating area, the number of workers required, qualifications for the job, and further, how this lack of herders would affect their production goals and further the food production program. After any rancher submitted an affidavit to request a former sheepherder to go back to the ranch, the worker was advised to get released with instructions for him. On April 19, Pete Etcheverry, a Basque sheepman who teamed with Bert Arambel at the Alpine Ranch, addressed John Dangberg, explaining in his broken English their own personal situation in Fallon. Etcheverry said that they had already applied for the release from the army of the following two sheepherders who used to work for them: Pierre

Goyhenetche, who was drafted into the army one year ago; and Jean Goyhenetche, who worked for five years at the Arambel & Etcheverry company. Etcheverry noted that they did not have any results and were "desperately in need of men." Previously, Etcheverry had further contacted Senator McCarran and asked for his help in getting relief for the agricultural labor shortage.[18]

In late April, after the office of the Farm Security Administration in San Francisco had received the application of the one hundred herders from Mexico, then, the Department of Agriculture certified the need of those workers through the Immigration and Naturalization Service. By early May, the Immigration Service's officials in Mexico City undertook all steps to recruit workers in Mexico to send at the very earliest opportunity on to Nevada. The Bracero Program of 1942, however, impeded the recruitment of those Basque immigrants living in Mexico. This bilateral labor agreement, which had been signed between the United States government and Mexico to relieve the unemployment problem in Mexico and the agricultural labor shortage in the United States,[19] was confined to the recruitment of Mexican citizens and therefore excluded noncitizens, particularly recently arrived Spanish refugees. Still, Nevada ranchers were confident that Basques in Mexico, not yet citizens of Mexico, could make their way into the United States legally under the Bracero Program. Contrary to those expectations, however, U.S. immigrations officials notified Grover B. Hill, assistant secretary of the Department of Agriculture, about the impossibility of recruiting Basques given this migrant labor program and remarked that only Mexican nationals were eligible for entering sheepherding jobs in Nevada and the West. On May 5, Grover Hill gave McCarran the bad news that "Basques cannot be recruited under the present agreement."[20]

Two days later, McCarran wrote John Dangberg to inform him about those legal problems of obtaining Basque sheepherders from Mexico. However, the stubborn Nevada senator refused to give up and declared that he was going to turn all his efforts toward securing Basque immigrants for Nevada because, in his own words, "they are almost always much more effective in this type of work." On May 10, then, McCarran contacted Earl G. Harrison, commissioner of the Immigration and Naturalization Service in Philadelphia, regarding the importation of Basque immigrant workers from Spain. Harrison recommended the following steps: first, file an application for permission to bring

Basque immigrants with the District Director of the Immigration and Naturalization Service in Denver; second, file an application with the local office of the Manpower Commission, with a view toward securing certification from this agency to the unavailability of labor of the kind that is desired that is currently unemployed in the United States. The NRSOA had already followed these steps. Harrison's words did not calm McCarran, nor the Nevada sheep ranchers.[21]

The barriers to recruit Basques from Mexico pushed both the NRSOA and the ENSGA to try to find these immigrants in other Latin American countries. On May 27, Basque immigrant communities were spread over a number of countries in South America, such as Chile, Cuba, Venezuela, and Argentina. The Basque immigrant community of Nevada maintained close ties with their families in the Old World. During wartime, they closely followed the events affecting their homeland. Oftentimes, members of the Basque community collaborated with nearby woolgrower associations, providing names of individuals and detailed information on both overseas workers seeking to escape Francisco Franco's Spanish regime and refugees who had resettled in Latin America. In more informal ways, Basque Americans involved in ranching supplied names and personal information—typically, relatives or friends from the same hometown—willing to fill the vacancy posts in the sheep industry. These efforts to bring Basques to Nevada promoted a sense of solidarity among the members of the Basque immigrant community.[22]

Yet, U.S. immigration law severely restricted Basque immigration from either Spain or any other Latin American country. There was, however, some hope. The Section 5 of the 1885 Alien Contract Labor Law provided some exceptions for the importation of immigrants, including the permission of foreign "skilled" workers, if labor of such kind could not be found in the United States. Under waiver of this excluding provision of the Alien Contract Labor Law, ranchers desiring to recruit those Basque immigrants, either from South America or Spain, had to personally request them through submission of a labor certification application. Then, Nevada sheepmen could fill out applications for permission to import Basque workers in the local offices of both the Immigration and Naturalization Service and the War Manpower Commission. Simultaneously, these sheep owners could still apply for Mexican workers at the offices of the Farm Security Administration.

The Farm Security Administration assisted those sheep operators interested in bringing Mexican workers. There were no federal programs, however, to assist those ranchers desiring to import Basque refugees from South America. The ranchers trying to bring labor from South America, according to McCarran, "were on their own hook." McCarran complained that federal agencies could not assist in the importation of this type of labor from South America and concluded, "It seems just as desirable to me to have Federal assistance in importing Basque aliens, as it is to have the Farm Security Administration assist in the importation of Mexicans."[23]

Nevertheless, the Basque labor shortage continued, causing a big unrest among the sheep community all over the American West. They continuously blamed the federal government for restrictions against immigration and also the wage differences between old and new industries in the West. Ranchers complained that new emerging private sectors were stealing their sheepherders through unfair wage differences. Although they raised wages to make sheepherding jobs attractive, most of the former herders still preferred working in other industrial sectors emerging in the West. As the labor shortage seemed to have no solution, ranchers continued arguing that Basques were the only skilled workers qualified for sheep grazing.

A Paradiplomatic Adventure: Martin Goñi and the Seventy-Four Basques from Navarre

On May 28, 1943, the western Nevada sheepmen from the NRSOA requested permission to import seventy-five Basques from either Spain or France. On June 3, Senator McCarran convened an urgent meeting with both Grover B. Hill of the Department of Agriculture and Earl G. Harrison of the Immigration and Naturalization Service to "the end that some definite results be forthcoming at once." Since the agreements between the United States and Mexico prevented the importation of Basques from Mexico, McCarran had put on the table the proposal to recruit Basques from other Latin American countries, but there had been little success with that effort. The bureaucratic process of requesting Basque immigrant workers had proven slow and long for the sheepmen.[24]

As part of the request submitted on May 28, on June 30, John Dangberg and D.W. Park, on behalf of the NRSOA members, submitted an affidavit in the Douglas County courthouse to request the importation of seventy-five Basque sheepherders. Then, they sent an application for that purpose to the Immigration and Naturalization Service office in San Francisco. Dangberg and Park prepared a long affidavit justifying that the Basque immigrant workers were the most dependable variables in the financial equation of the sheep industry. To put it another way, according to their statement, the annual production would be curtailed, unless Basque labor was forthcoming. They pointed out, "the range sheep industry has been built on the peculiar skill and knowledge of French and Spanish Basques, that this organization can now place seventy-five such men aged twenty to forty years as sheepherders." They went further by describing Basques as naturally adapted to sheep herding. They wrote that "the skill of [Basques] rests on their having learned in their native Pyrenees from early childhood to understand and observe the habits of sheep and goats on high mountain ranges," among other things. Besides their alleged herding qualifications, Dangberg and Park said that since the early twentieth century Basques not only became the backbone in the labor pool of the sheep industry, but also they had consolidated as a legitimate immigrant community in Nevada.[25]

But while they enthusiastically glorified and praised Basque sheepherders, the request would again discriminate against Mexicans as undesired labor. "For the most part Mexicans are dissatisfied in the area owing to the fact, apparently, that there are no Mexican communities to which they can go for recreation and also to the fact that winter at these latitudes is too cold for them." Dangberg and Park explained as follows the allegedly ideal labor environment that existed among the members of the Basque sheepherding community in Nevada: "Affiants, each for himself, states that to the best of his knowledge there has never been any strike or other concerted action of herders against their employers and each further affirms that he has never heard of any concerted action of employers against herders and that the employer-employee relation has always been personal and individual." Since herders spent long periods of time alone far from the employers' supervision, docility was one of the most desirable properties of labor in the sheep industry. In addition, they declared that the herders' wages had gone up by 71 percent during World War II—from about $70 to $120. They concluded that if there was not early relief from the Basque labor shortage, then it would result in the final "reduction of flock or complete liquidation of them . . .

[and] such liquidation will lead to the loss to the nation of wealth in our mountain areas." All in all, thus, the NRSOA requested seventy-five Basques from the Old Country as soon as possible in order to work with the sheep in Nevada.[26]

On the same day, June 30, John Dangberg wrote letters to the local offices of both the Immigration and Naturalization Service and the Employment Service in Reno, notifying them of the affidavit submitted in the request for the seventy-five Basque workers. He also wrote to McCarran noting their application. Dangberg stated that this application was only a small patch within the larger problem and suggested that the situation required an exceptional measure. He said that the 1924 immigration law was strangling woolgrowers because it impeded the recruitment of more Basque immigrants. He stated the difficulties associated with "the situation with respect to slaughter quotas and the highly dangerous effect" on sheep production. He continued by saying that feed was very short and "the quota system is tending to back up finished stuff in the hands of producers thereby reducing the available supply of vital feeds for finishing livestock later in the year." McCarran for his part agreed with Dangberg and blamed the frequent administrative transfers between different government agencies as holding up the procedures to import Basque immigrants. He criticized the federal bureaucracy as follows: "I got so 'damned' mad at the inefficient, mismanaged bungling of this job in the various federal agencies that I nearly popped. . . . I might say that I necessarily refrained from mentioning a number of things that were running through my mind." Although frustrated, McCarran continued to seek solutions that would enable the importation of Basque immigrant workers.[27]

During that summer, while the American and British troops were gaining ground in Italy as part of the Allied military offensive against the Axis powers, the broader ranch labor situation did not significantly improve, but it was slightly better than last summer largely because of the concerted efforts of the sheep operators. However, this did not reduce the broad effort to recruit Basque laborers. On July 15, Colonel Philip G. Bruton, deputy of the War Food Administration's corps of engineers, informed McCarran that all the attempts to induce the Mexican government to allow the importation of Basque refugees who were residing in Mexico had failed. Considering the acute situation and the refusal of the sheepmen to employ Mexicans, Bruton informed the senator about the availability of experienced "Spanish Americans" in New Mexico. He suggested that hiring those sheepherders could work as an alternative to the Basques. Bruton explained, "These people do not have as good a background as the Basques, but if we do not have any success in importing Basques, we are going to have to develop the next best source."[28]

McCarran, however, was not at all satisfied with Bruton's answer and appeared very skeptical at his suggestion: "I wish to say right now that I do not know whether these Spanish-Americans can be so trained as to become good sheepherders," he wrote. The Nevada ranch owners were also not pleased with the idea. According to John Dangberg, although the recruitment of those proposed "Spanish-American" workers in New Mexico could help to answer the herder shortage, "I am not sure that I know what these people are, whether South Americans or naturalized Spaniards." Having in mind the abundant Mexican agricultural labor in the Southwest, Dangberg questioned the qualifications of those sheepherders along the lines of a racial and ethnic prejudice. Nevertheless, he wanted to stay optimistic about the possibility of recruiting these workers and further said that "anyone who is *willing* [*sic*] to herd or to learn to herd will be better than no one at all." However, he noted, "the employment of them might not be a very workable thing." The Forest Service proposed that these laborers be sent to sheepherding training programs. To Dangberg, this option was impractical because there was no time to properly train unskilled laborers—according to him, this required at least three years. For these reasons, Dangberg reaffirmed his idea that their "best hope lies in the direction of Spanish or French Basques if and when the international situation clears up."[29]

In other western states, while many sheepmen were having considerable difficulties in securing Basque sheep labor, they started to employ Mexicans. Nonetheless, Dangberg and other Nevadans persisted to find ways to bring Basques. However, in those years after the Spanish Civil War, not only did immigration restrictions present obstacles but Nevada ranchers experienced additional difficulties because of Francisco Franco's rule and his support of the Axis powers against the Allies during World War II. In the fall of 1943, the recruiting process of Basque herders from Spain was further interrupted by the war events in the Mediterranean.[30]

Francisco Franco's nonbelligerent stance, yet decidedly pro-Axis sympathies, made Spain a "thorny issue" for the Allies. In the summer of 1943, the United States and Great Britain launched an offensive in northern Africa, aiming next at Sicily and then the Italian peninsula.

Since the outbreak of the Spanish Civil War, the Roosevelt administration's foreign policy toward Spain was marked by a profound ambivalence and improvisation. When the United States formally entered the war after the bombing of Pearl Harbor on December 8, 1941, Franco, for his part, adopted an increasingly opportunistic and pragmatic policy to prevent any possible attack by the Allies. The ideological implications of Franco's regime and its collaboration with Adolf Hitler disturbed the uneasy relations between the United States and Spain. In close cooperation with Great Britain, the U.S. government had aimed to prevent Franco from joining the Axis powers and to keep the fascist regime neutral by supplying Spain with food goods and other raw materials, especially petroleum, needed to keep Spain's ruined economy afloat. At the beginning of the war, Roosevelt had approached Franco with the aim of distancing the dictator from the Axis bloc. Later on, Roosevelt further warned Franco that his collaboration with Hitler would have serious consequences in its future relationships with the United States. Under pressure from a domestic public opinion strongly against the Spanish fascist cause, the Roosevelt administration pursued a careful policy, which consisted of pumping supplies into Spain in order to keep Franco away from the war. As long as Franco cooperated with Hitler, the U.S. government did not consider any formal agreement with Spain. Nevertheless, in Spain, Franco's regime maintained its international alignment with the Axis until the demise of Hitler's Germany.[31]

After the summer of 1943, and despite the course of the war against the Axis powers in southern Europe, the Immigration and Naturalization Service in San Francisco started processing the application for the importation of the seventy-five Basques. The recruitment of this group of Basque workers had to be admitted under the terms prescribed in the provision applying to the importation of "skilled" contract labor in the Immigration Act of 1917. In order to determine the application of the law, the Immigration and Naturalization Service in San Francisco requested that the NRSOA provide exact locations of the seventy-five Basques whom the Nevada ranch owners desired to employ. The NRSOA, however, was unable to provide exact addresses of Basques in Spain because many of the Basque Americans living in Nevada had lost many contacts since the civil war broke out and the subsequent establishment of Franco's rule. However, Dangberg was optimistic and thought that, in a near future, "it would be possible to reestablish relations with these people and again recruit qualified herders from the area that has supplied the skilled men on whom the range sheep industry depends for existence."[32]

Once again, Dangberg made it perfectly clear that the selection for the recruitment of sheepherders was to be restricted to Basques. Interestingly enough, Dangberg suggested to immigration officials how the importation of Basque sheepherders might occur in such a critical international context. He proposed to send immigration agents to the main Basque ports—Bilbao and San Sebastián—as well as other small coastal towns. Unemployed men might enroll as a way to escape from Franco's Spain. From there, they would embark onboard any ship, which periodically sailed from South America to Spain. Following Dangberg's suggestions, then, those Basque immigrants would endure an arduous triangular transatlantic journey: sail from Spain, pass through any South American country, and eventually deliver themselves in any state in the American West.[33]

Furthermore, Dangberg proposed for the first time to the Immigration and Naturalization Service to use the issue of the Basque immigrant sheepherders as a bargaining chip in the foreign relations between the United States and Spain. Dangberg wrote in a letter to the Immigration and Naturalization Service in San Francisco dated October 5: "It would seem from other information which we now have that the Franco government might be willing to exchange herders (Basque citizens) for supplies." In another letter of October 23, replying to the Immigration and Naturalization Service's inquiries of eligible sheep workers' contacts, Dangberg pointed out again the conditions of relations with Spain: "From what we read, it seems to us that the present policy of our Government is to maintain good relations with the existing Spanish Government, and that it is the strong desire of the Spanish Government to maintain good relations with both the United States and Great Britain." Interested in recruiting Basques by all means, Dangberg suggested that the sheepherder question could serve as an indirect vehicle to conclude broader strategic understandings between the United States and Franco's Spain, despite the differences that prevented the fulfillment of a mutual agreement between the two governments. Dangberg forwarded this correspondence to McCarran. In response, the Senator from Nevada quickly decided that in order to secure sheep laborers from the Basque Country, dialogue with the dictator Franco must happen.[34]

In November 1943, the Immigration and Naturalization Service granted the NRSOA permission to import seventy-five Basque immigrant workers for temporary employment by the members of the woolgrower associations for a period ending October 15, 1944. They required employers to post a bond in the sum of $5,000 in order to ensure departure of the workers within the time period of the visa, and required a forfeiture sum of $500 for any failure to depart. But this temporary permission to import the seventy-five Basques was not well received at all since these immigrants had only one-year work permits. John Dangberg persuaded Thomas B. Shoemaker, deputy commissioner of the Immigration and Naturalization Service, to extend the time permitted to a period of three to five years. According to Dangberg, one year was not enough time for an immigrant to adapt to a new work environment. However, the Immigration and Naturalization Service was not in a position to grant permission for the importation of those Basque immigrants for a longer period than one year. But ranchers still could make an application before the expiration date for work permit extension.[35]

The Immigration and Naturalization Service then informed the State Department's Division of European Affairs in order to discuss the importation procedure of the seventy-five Basque herders from Franco's Spain. The Division of European Affairs suggested that as far as transportation was concerned, there was no need to first send Basque immigrants through any South American ship, as Dangberg had previously suggested, because it was already operating the longstanding *Compañía Transatlántica Española* [Spanish Transatlantic Company], founded in 1850, which plied regularly between the North Atlantic ports of Spain and the port of New Orleans. Second, regarding the issue of reaching an international agreement in order to export Basques from Spain—because this transatlantic passenger ship business was already active in Spain—the Division of European Affairs contended that "there would appear to be no necessity for any diplomatic action in this regard." Nevertheless, the State Department indicated that those Basques desiring to enter the United States had to make an application individually in the usual manner for immigration visas.[36] Although the American consular officers abroad were eager to collaborate with the NRSOA undertaking the recruitment of the seventy-five Basques from Spain, the US–Spanish relations were still uncertain.[37]

This transatlantic traffic flow referenced by the Division of European Affairs reinforces the idea that by that time, Franco's regime was not as hermetic and isolated as has traditionally been said. This case also demonstrates how the State Department's approach to Franco's Spain differed from the U.S. administration's during World War II. While the State Department was more prone to negotiate with Franco and strengthen commercial ties between the two countries, Roosevelt disagreed over whether to continue military and humanitarian aid to Spain. In any case, however, Franco was clearly moving toward a warmer relationship with the United States. In late 1943, there was no doubt of an Allied victory. In October, Franco had reaffirmed its position of "strict neutrality" in the war, and as the year ended, Franco's regime progressively began to move toward the Allies and the United States.[38]

On December 4, the NRSOA informed its members about the latest progress in the effort to get Basque sheep workers into the United States to replace the old and ill herders, as well as those men who transferred to defense industries and other jobs. The head of NRSOA also briefly explained the steps involved in importing Basque immigrants: first, members of the association individually must contact eligible Basque workers and arrange means of their transportation; second, those contacted individuals must appear before the American Consulate office in Bilbao to apply for visitors' visas to enter the United States; and third, the head of NRSOA also guaranteed its members that the association would provide assistance in any way possible to obtain extensions of time for those herders to remain in the United States. The American Embassy in Spain decentralized this recruiting-based immigrant visa process to the consular office in Bilbao, Biscay, in order to both facilitate Basque enrollment and concentrate application processing in the Basque Country.[39]

After a long and intense process, the U.S. government finally allowed the recruitment of Basque immigrant workers for employment in Nevada. On December 18, McCarran wrote an encouraging letter to Dangberg saying that they had already laid the foundation for subsequent importations of Basques from the same country: "We had a long hard fight on this matter, but I believe the groundwork has been so laid that additional importation of Basque herders will be handled more expeditiously." For their part, Nevada ranchers celebrated the results obtained and other western woolgrowers congratulated Dangberg for his work. However, the NRSOA still had some work to do before the recruitment of those Basques could be fully accomplished.[40]

The NRSOA needed to advertise the existence of these jobs in the Basque Country. Otherwise, no Basque would have found out about the possibility to emigrate and become a sheepherder in the Far West under the Francoist regime. It became increasingly difficult to recruit Basques from traditional sources because of the postwar problems in Spain and the context of war in Western Europe. Dangberg had requested help from the Basque members of the NRSOA to aid in the process of informing their countrymen about the possibility of emigrating by applying for job openings. Nonetheless, the Basques of Nevada could not contact their family members and friends in such a turbulent period. Also, the Basque-American community doubted that the Spanish state would allow the recruitment of Basques, as Franco's repressive apparatus had particularly targeted this population. Indeed, contrary to the State Department's optimism, Francisco Franco's rule posed new challenges for importation of this labor. Before continuing with the selecting and hiring process, the NRSOA wanted to have assurance from Franco's regime that it would permit the recruitment of workers from the Basque Country.[41]

During the early years of the dictatorship, Franco's regime did not have a well-articulated foreign policy. Neither did it have a clearly defined plan of action on certain kinds of international issues, such as the emigration of Basque laborers. In August 1941, Franco's regime had prohibited by executive order emigration from Spain. This emigration restriction served as another repressive act against the defeated Republican army, opponents, and those considered enemies of Franco's regime who were interested in leaving the country. At first, however, the regime could not strictly enforce the prohibition of emigration due largely to institutional weaknesses and lack of expertise on these issues. In the early 1940s, after the Spanish Civil War, Franco's systematic repression of political opposition, increased poverty levels, and prolonged food shortages pushed many Basques to go into exile. Large groups of exiles and a rural-urban migration process caused significant population decrease and displacement. As a result, in the early postwar years, with World War II as a backdrop, Franco's regime needed cheap and pliable labor for the reconstruction of the Spanish economy. Consequently, any exportation of labor from Spain to the United States, such as the Basque immigrant workers, would have to be greatly compensated.[42]

By the end of 1943, with Allied victories, Franco had adopted an increasingly pragmatic policy reaffirming not just nonbelligerence but now neutrality. Taking advantage of this favorable moment to open negotiation, on December 28, McCarran, responding to the supplications of Dangberg, addressed Juan Francisco de Cárdenas, Spanish Ambassador in Washington, D.C., outlining the situation to him and urging cooperation with the United States in this matter. McCarran asked permission to bring seventy-five workers to the United States for sheepherding and to send a licensed Basque American from Nevada to undertake the recruitment work in the Basque Country.[43]

On the next day, Cárdenas informed McCarran that the present law prohibited Spanish labor recruitment in foreign countries. Cárdenas invoked the 1907 Emigration Law, which read as follows: "The recruitment of immigrants and propaganda to foster emigration, as well as the existence of agencies directed to that purpose, are prohibited, under penalties established by law." In 1907, the Spanish government of Antonio Maura had passed this emigration law as part of its conservative "top-down revolution" that looked at more government intervention and regulation to address the socioeconomic problems affecting Spain in the early twentieth century. Owing to these legal impediments to recruit Basque workers, Cárdenas explained to McCarran that he could not do anything to help with this matter before obtaining central governmental authorization. Cárdenas suggested McCarran contact the U.S. Embassy in Madrid through the Department of State as a way to approach Franco's regime and for "explaining the case to see whether, in view of the situation created by the war, something could be done." This correspondence marks the beginning of the negotiating process between Senator McCarran and Franco's regime that would lead to other diplomatic efforts beyond the recruitment of Basque immigrant workers.[44]

On January 3, 1944, McCarran informed Dangberg that "the Spanish immigration laws are dead against us and therefore we will have to continue to work through the State Department." After receiving McCarran's notification and a generalized sense of fear among the local Basque immigrant community, Dangberg came to realize that it was too risky sending a Basque American from Nevada into Spain to recruit workers who would have been under the threat of Franco's military. The NRSOA considered the possibility of making contacts directly through a middleman living in Spain to arrange all the recruiting process. However, considering the character of the Francoist regime, Dangberg suggested that they develop a diplomatic means through the

U.S. Embassy in Madrid to ensure completely safe arrangements: "We believe, however, that we will be in a better position and the men will also if as the Spanish Ambassador suggests we take up the matter with our Embassy in Madrid to see whether, in view of the situation created by the war, something could be done." In early 1944, the United States increasingly pressured Franco to meet its demands of strict neutrality. After continued diplomatic negotiations with the U.S. government, Franco yielded to the Allies' pressure to adopt an increasingly neutral policy and completely eliminate Spanish assistance to Hitler (even though it was small), such as eliminating wolfram exports to Germany or interrupting the logistical support to the Nazi espionage activities in the Spanish territory. Dangberg viewed these political changes as a good opportunity to seek approval of Franco's regime at the earliest possible date to proceed with the recruiting process in Spain.[45]

The NRSOA continued working toward their goal of once and for all bringing the group of Basques to Nevada. In late January 1944, John Dangberg hired a Basque in Spain to work for them as a labor agent in order to search for workers and recruit a labor crew. The man in question was Martin Goñi, a former sheepherder in Nevada known by many ranchers who had been living in the Basque Country in his hometown Urritza, Navarre, for ten years after previously working in Nevada for many years. Goñi accepted the job offer to recruit Basque laborers and channel them to Nevada, acting as the middleman. In this position, he became the intermediary not only between the NRSOA and workers, but also between Spanish and U.S. immigration-related authorities.

By early February, Martin Goñi had already recruited a crew of men from the province of Navarre ready to depart to the United States. Goñi said that, if all of the required procedures were completed properly, these Basque immigrant workers could arrive in Nevada by mid-June of the same year. Ranchers were required to follow the following procedures in order to secure these workers: first, send a check for steamship fare in the sum of $428 per worker; second, pay for incidental costs, including Spanish and U.S. taxes, amounting to no more than $50 for each man; third, as requested by the U.S. government, file a bond in the sum of $500 for each worker imported, guaranteeing his return-trip to Spain; and forth, advance the rail fare from New Orleans to Reno. Furthermore, the NRSOA had taken out a travel insurance to cover unexpected problems while traveling from the port of Bilbao to Nevada.[46]

By late February, the NRSOA had collected all the money for the issuance of visas and arrangements to purchase tickets for the prospective immigrant workers. But then, a new problem emerged. The post–Civil War context in Spain, marked by severe repression and oppression, posed a threat to this transaction and to the agent Martin Goñi. On March 17, 1944, in view of the money license exchange difficulties, Howard K. Travers, chief of the State Department's Visa Division, recommended Dangberg to deposit transportation and visa costs with the New York office of the *Compañía Transatlántica Española*, the steamship company concerned. However, the fares paid in the United States doubled those paid in Spain. Dangberg sent a telegram asking if it was possible to handle the "advance of fares" through the U.S. Consul in Bilbao, Harry F. Hawley. Dangberg presented Travers with the idea of inducing the American Consulate in Bilbao to make the payment of funds to Martin Goñi or even supervise such a payment, considering the complicated context in Spain at this time. Eventually, this idea was pursued, and the NRSOA made the transaction of $30,000, ensuring that the transaction was complete with all guarantees given the instability after the war in Spain. By that time, Cordell Hull, secretary of state, had become directly involved in the Basque labor issue himself. In the shadows of the war, while the United States was negotiating with Franco from a superior economic and military position, the Basque labor matter went so far as to involve even the highest levels of the State Department.[47]

On March 23, 1944, Francisco de Amat, Spanish Consul in San Francisco, sent a telegram to the civil governor of Navarre certifying the NRSOA through its labor agent Martin Goñi to recruit a group of Basque workers and send them to Nevada. At this moment, Franco's regime accepted the conditions to export the Basque crew to the United States. However, Spanish authorities also counted Martin Goñi within the total of seventy-five Basques permitted to emigrate. Thus, Goñi had to recruit seventy-four Basque immigrants. Surprisingly enough, Francisco de Amat required the NRSOA to sign a labor contract that aimed to define decent working conditions for the Basques, such as: prevailing wage, protection in case of injury during employment, and even guaranteed return fares if they were not satisfied. While Dangberg progressively completed all the formal steps to bring the seventy-four Basque herders from Spain and pursue diplomatic channels to address the Basque labor problem, John K. Robbins and D.A. Hughes, president and secretary of the ENSGA respectively, were in Mexico in search of Basque herders. At the same time, other western woolgrowers associations were looking to the Nevada sheepmen's associations as the leaders in monitoring and recruiting Basque immigrant workers. As a new lambing season was approaching and the unsolved labor shortage hampered sheep production, frustration and worry continued to spread across all the American West.[48]

By mid-May, after eighteen months of intense spadework, Goñi had everything set up for sending the Basque workers to Nevada. The crew was divided into two groups and departed separately to the United States. On May 14, 1944, the first group consisting of forty-two Basque workers, including Goñi who was in charge of the crew, departed from the port of Bilbao. From Bilbao to New Orleans, these Basque immigrants endured a long voyage on board the steamship *Marqués de Comillas* of the Spanish Transatlantic Company, stopping first in a Cuban port. Meanwhile, Dangberg made all the necessary arrangements for the reception of the Basque crew with the Immigration and Naturalization office in New Orleans. Also, he purchased the tickets for the Basque immigrants from the Southern Pacific Lines of the New Orleans & Texas Pacific Railway Co. and made all the preparations with A.C. Ziegan, division passenger agent of the same company in New Orleans, to bring them to Nevada. At the same time, Dangberg continued with the legal steps to secure the remaining thirty-two Basque immigrant workers who sailed from Bilbao on June 15.

One month after the departure, on the night of June 16, the first group of Basque workers arrived at the port of New Orleans. The next morning, after passing through immigration control, Martin Goñi and the forty-two other Basques embarked on a train ride from New Orleans to Reno, where they arrived five days later on June 22, set "to become sheepherders on Nevada ranges," as the *Reno Evening Gazette* reported. The editorial of the *Gazette* described the Basques' arrival as follows: "The Basque contingent arrived shortly after 9 AM and the men left the cars, many of them wearing the characteristic small black berets." In the train station, the impatient ranchers and other members of the local Basque-American community greeted the immigrants. Immediately after their arrival, they went to the Santa Fe Hotel, one of the local Basque boardinghouses, to have breakfast. Afterward, workers and employers met with Harmon E. Hosier, the officer in charge of the Immigration and Naturalization Service, to complete all necessary legal paperwork. Considered as a highly valuable human commodity, the names of the Basque immigrant workers were written on slips of paper and placed in a hat. Ranchers put their hands into the hat and pulled out slips. The names of the individuals written on the slips became the ranchers' new employees. Soon after the raffle, ranchers went back to their respective ranges with their newly hired workers. One month later, the remaining thirty-two Basque workers reached Nevada. After tremendous effort to bring these workers from Spain to work in Nevada, the recruitment of the seventy-four (plus Martin Goñi, seventy-five) Basques did alleviate some of the labor shortage in Nevada. Nevertheless, it did not solve the entire labor shortage problem in the American West.[49]

Some days later, Dangberg wrote that the seventy-four Basque immigrants "seem to be sober and industrious young men." Dangberg had indirectly justified the exclusive recruitment of Basque immigrants on the grounds of their superior skills. According to him, Basques had racial attributes that prepared them for sheepherding. Against this argument, however, most of those Basques had never worked before in the sheep industry or were not qualified herders. One of those seventy-four Basque newcomers, José Antonio Santesteban, who arrived in Reno in the second group, wrote a letter in August in Spanish to Dangberg requesting a transfer to another job because he had no idea how to work with sheep. On his arrival with the rest of the crew in Reno, Santesteban was assigned to work for the ranch owned by the Basque sheepman Bert Ithurburu in Susanville, California. Santesteban desired to change his occupation as a sheepherder because, as he pointed it out to Dangberg, he had no experience working with livestock: "I have never worked in this type of occupation in my life and I don't know anything about raising livestock." In this letter, Santesteban explained that before his recruitment, he made this clear to Martin Goñi and also that he could work as a mechanic. Goñi overlooked Santesteban's inexperience with sheep work. This was not a problem and he included Santesteban in the work crew. Santesteban simply desired to emigrate to escape Franco's regime, find a new life, and once settled in the United States, send for his wife. This testimony reinforces the thesis that Basque immigrants were not by nature adapted to sheep grazing or even skilled herders as Dangberg and other ranchers continuously insisted in their quest to employ Basques.[50]

"Keep It Working": Discretion and Cooperation

The case of the seventy-four Basque workers represented a political success for Nevada. Sheepmen all over the American West had great admiration for the Nevada ranchers. Further, other woolgrower associations contacted Dangberg for guidance on importing more Basques from Spain. Besides advising these woolgrowers, Dangberg urged them to keep working with their representatives and encouraged more effective cooperation among western associations on the importation of the Basques. Dangberg suggested other secretaries of woolgrower associations "explore the future," working discreetly without attracting excessive attention. On August 7, 1944, Dangberg noted: "Publicity is not desirable and may have serious repercussions if it attracts the attention of organized groups, leading to agitation and conflict which would, in turn, create an unfavorable attitude towards importation of men on the part of the government agencies involved." In view of the urgent need for more labor, according to Dangberg, western woolgrowers had to act together and take advantage of the favorable circumstances to set up formal programs for more Basque immigrant workers on a regular basis. In the fall of 1944, there was closer collaboration among western woolgrower associations.[51]

John Dangberg saw it necessary to reach a stable diplomatic agreement that would allow recruiting and exporting Basques to the United States. Since Franco's regime prohibited private employment agencies from operating in Spain, Dangberg proposed making these arrangements in more informal ways. Intending to evade the Spanish legislation, he suggested working through a Basque man "acting in response to a request of the boys and *not* [*sic*] as the agent of the importing agency." McCarran also did not give up on the matter. On August 24, McCarran wrote to Dangberg: "After the difficulty in setting up the program, we must do everything possible to keep it working." McCarran took this issue up through the State Department and the American Embassy in Madrid. On September 7, 1944, Secretary of State Cordell Hull wrote a letter to Pat McCarran expressing his willingness to cooperate and reach an agreement to appoint an agent in Spain to assist in the recruitment of the Basque immigrant workers. The State Department transmitted all the correspondence of McCarran and Dangberg to the U.S. Embassy in Madrid.[52]

Other western woolgrower associations also worked to import Basque workers from Spain emulating the work of the NRSOA. On January 1, 1945, Dangberg wrote a long letter to M.C. Claar, secretary of the Idaho Wool Growers Association, in which he encouraged him to keep working on the importation of more Basque immigrant workers, but remained honest about the difficulties of this adventure. In order to achieve successful results, as Dangberg explained, it was necessary to have close cooperation with their congressional delegation and local ranchers' strong commitment to this matter and be ready to "make the decisions and assume responsibility when unexpected situations arise." Dangberg wrote: "By far-seeing, we mean men who will see that their own labor difficulties can only be solved if they work out a solution that benefits the entire area or state participating." In January 7, 1945, six days later, the Idaho Woolgrowers celebrated their Fifty-Second Annual Meeting, in which the Basque labor shortage was the first major point to discuss.[53]

From California, woolgrowers also began making plans to import a group of Basques from Spain. In the month of January 1945, assisting his constituents' needs, California representative Alfred James Elliott addressed the Spanish Embassy in Washington, D.C., requesting permission to recruit Basques in Spain. On January 31, 1945, Eduardo María Danis, Spanish first consul in Washington, D.C., wrote back to Elliott saying that "neither this Representation nor the Spanish government is opposed in principle to the plan, though we feel that this is a question to be carefully tended to, if the interests of those Basques sheep herders are to be duly protected." Comparing with the previous years, furthermore, Elliott appeared more interested to reach an agreement and he suggested creating a guest worker program similar to the Mexican Bracero, but with the Spanish government:

> Former contracts concerning Spaniards brought to this country for the same purpose, have not been quite satisfactory and difficulties have arisen, which, if possible, we wish to avoid in the present case.
>
> An agreement between the Governments of Spain and the United States, similar to that existing between the United States

> and Mexico for the engagement of Mexican workers, would be the best solution, and the efforts of this Embassy will be directed to that purpose. But, since this is a question which will probably take some time to decide, and apparently the case you refer to has a certain urgency, it would perhaps be convenient to try to find a temporary arrangement.[54]

This approach to the recruitment of Basque sheepherders by Eduardo Danis reflected a change in Francoist foreign policy toward the United States. Although informal, closeness between the United States and Franco's Spain increased, as the Allies conquered more territory from Germany and demonstrated their military power.

Dangberg for his part opposed the suggestion of the Spanish Embassy to articulate a plan similar to the Bracero program for Basque sheepherders. Dangberg said that he preferred that Basque immigrant workers not be classified as temporary workers, but as permanent, because ranchers wanted long-term workers and perhaps permanent residents. Furthermore, Basque immigrants themselves asserted that they "*will not leave Spain* [*sic*] unless they can come with some assurance that they are coming for *permanent residence* [*sic*] or for five or ten years." He added:

> Believing that the need for the men would continue to be critical and that our government would recognize this fact, we gave them assurance that this would be the case. These Basques are individualists and one of their reasons for emigrating is to realize greater freedom in America. It does not seem that an intergovernmental program would give them this desired freedom even though it might, for a time, provide fares from Bilbao to California.[55]

Dangberg informed McCarran about his posture toward the proposal from the Spanish Embassy to make a program of Basque migrant workers. To Dangberg, it was necessary for another kind of labor program to secure the employment of Basque immigrant workers for longer terms in the sheep industry.[56]

While other western woolgrower associations started working to import small groups of Basques following Dangberg's recommendations, the NRSOA and ENSGA joined forces to import a new crew of Basque immigrants from Spain. On February 15, 1945, John Dangberg contacted Howard K. Travers, chief of the Visa Division in the State Department, to make arrangements to import a new group of Basques.

This time, Dangberg asked Travers about the possibilities of recruiting Basque workers under the quota applied to Spain through a Basque middleman named Francisco Catalan not officially employed for recruiting labor. In this way, Dangberg aimed to evade Franco's legislation, which did not permit any labor agency to operate in Spain. Much like in the previous case, for the purpose of accepting Basque immigrants as permanent residents, as distinguished from the rules applied to temporary workers, the U.S. government required Nevada sheepmen to apply to the Immigration and Naturalization Service for a waiver under the contract labor clause of the Immigration Act of 1917. This meant entry could only occur if the workers fit into a category of "skilled" labor that was not available in the U.S. population. Then, Nevada woolgrowers had to file immigrant visa petitions in the United States to employ more Basque immigrants from Spain which at the same time were subjected to the quota limits. Once the Immigration and Naturalization Service approved these individual applications, the State Department, through its consular office in Bilbao, proceeded to arrange the paperwork to allow these Basque immigrants enter the United States. Dangberg and other ranchers were frustrated with the slow procedure for determining immigrants' eligibility for this particular visa. Thus, Dangberg suggested the need to establish an exceptional plan that would allow employing Basque immigrant workers from time to time responding to the ranchers' needs.[57]

At the end of spring 1945, the war in Europe changed in important ways. In May 1945, Nazi Germany surrendered unconditionally to the Allies. After the fall of Nazism, during the summer months, the war front moved to the Pacific. During the final stages, following the United Nations Conference on International Organization in San Francisco, on August 2, 1945, President Harry S. Truman, the Soviet Union's leader Joseph Stalin, and British prime minister Clement R. Attlee met in the Potsdam Conference where they tried to settle the international system to reorganize the ruined Germany and secure a peace agreement for postwar reconstruction. They also issued a joint declaration about the "Spanish Question," rejecting publicly Franco's regime and opposing its inclusion into the United Nations. Franco's Spain was internationally ostracized for its support of the Axis powers, being excluded from the American economic aid known widely as the Marshall Plan. After the collapse of the Third Reich, Franco's Spain stood alone before the rest of liberal western democracies and the Soviet Union. In these first years of the Cold War, Francoist foreign policy would be characterized by patience and pause: wait until the international context favored diplomatic ties with the United States.[58]

The lobbying efforts of Patrick McCarran not only satisfied Nevada woolgrowers' demands, but his work also became crucial in setting the stage for subsequent programs to bring more Basque workers from Spain. At the beginning of 1945, John Dangberg had evaluated McCarran's work positively and looked to an encouraging future on this matter under his patronage. He wrote: "We have been ably and sympathetically supported throughout by Senator McCarran, who has, in fact, fathered this enterprise. He has repeatedly rescued it from the bogs of bureaucracy and has laid the foundation for others to follow in our footsteps." By the beginning of the Cold War, the new post-war international situation, marked by a growing tension and hostility between the United States and the Soviet Union, channeled and legitimized the bilateral relations between the United States and Franco's Spain. The question of Basque immigrant workers served to open up dialogue and even closer relations between the two countries in the developing Cold War context.[59]

Summary

During wartime years, the work of the NRSOA with Pat McCarran became an example for other western woolgrower associations that confronted the same labor shortage. Furthermore, the commodity characterized as "skilled labor" was necessary when the NRSOA and McCarran had to justify the urgent necessity of admitting Basque immigrants. This attitude in turn reinforced the notions of Mexican otherness and backwardness in the sheep industry of the American West.

The federal agencies involved in the recruiting process of immigrants in the United States had not been enthusiastic about cooperating in the recruitment of Basques and insisted on the possibility of Mexican labor through the Bracero program. The NRSOA and other western sheep associations' refusal to employ Mexican workers and the desire to hire only Basques made clear the ranchers' wish to perpetuate the preferred Basque racial group to labor in the sheep industry. Despite everything, their efforts succeeded in obtaining the admission of a group of seventy-four Basques. Although this small victory did not solve the entire labor shortage, it set a precedent for subsequent actions to admit more Basques to Nevada and other western states. The scarcity of Basque immigrant labor remained a central problem for Nevada and western woolgrowers.

The next chapter takes up an examination of how the postwar international interests of the United States and Spanish governments began to set an agenda open to the recruitment of Basque immigrant labor. More than ever, the recruitment of Basque immigrant labor to work in the western sheep industry became an important issue in the national definitions of both countries.

Notes

1. W.P. Wing to John Dangberg, June 9, 1943, Collection no. 13, Box 39, Dangberg Correspondence 1943, McCarran Papers.
2. David M. Kennedy, *Freedom from Fear: The American People in Depression and War, 1929–1945* (New York: Oxford University Press, 1999), 585–90; Homer Mooney to John Dangberg, January 21, 1943, Collection no. 13, Box 39, Dangberg Correspondence 1943, McCarran Papers.
3. Tom Jolly to John Dangberg, January 22, 1943, Collection no. 13, Box 39, Dangberg Correspondence 1943, McCarran Papers.
4. W.P. Wing to John Dangberg, January 28, 1943, Collection no. 13, Box 39, Dangberg Correspondence 1943, McCarran Papers.
5. D.A. Hughes to John Dangberg, January 29, 1943, Collection no. 13, Box 39, Dangberg Correspondence 1943, McCarran Papers.
6. "Price and Labor Top Wool Parley," *Times-News*, January 10, 1943, 2; Dan J. O'Loughlin to John Dangberg, January 29, 1943; "Request For Workers," Collection no. 13, Box 39, Dangberg Correspondence 1943, McCarran Papers.
7. John Dangberg to W.P. Wing, February 1, 1943, Collection no. 13, Box 39, Dangberg Correspondence 1943, McCarran Papers.
8. John Dangberg to Tom Jolly, February 3, 1943, Collection no. 13, Box 39, Dangberg Correspondence 1943, McCarran Papers.
9. John Dangberg to Dan O'Loughlin, February 12, 1943; John Dangberg to D.A. Hughes, February 12, 1943, Collection no. 13, Box 39, Dangberg Correspondence 1943, McCarran Papers.
10. Paul V. McNutt to Patrick McCarran, February 12, 1943, Collection no. 13, Box 39, Dangberg Correspondence 1943, McCarran Papers.
11. Dan O'Loughlin to John Dangberg, February 15, 1943, Collection no. 13, Box 39, Dangberg Correspondence 1943, McCarran Papers.
12. Eva Adams to D.W. Park, February 15, 1943, Collection no. 13, Box 39, Dangberg Correspondence 1943, McCarran Papers.
13. John Dangberg to Patrick McCarran, February 23, 1943, Collection no. 13, Box 39, Dangberg Correspondence 1943, McCarran Papers.
14. John Dangberg to Patrick McCarran, February 23, 1943; D.W. Park to Patrick

McCarran, March 15, 1943, Collection no. 13, Box 39, Dangberg Correspondence 1943, McCarran Papers.

15. D.A. Hughes to John Dangberg, February 27, 1943; John Dangberg to H.L. Englebright, March 15, 1943; D.W. Park to Patrick McCarran, March 15, 1943, Collection no. 13, Box 39, Dangberg Correspondence 1943, McCarran Papers.
16. Dan J. O'Loughlin to Frank Reed, March 24, 1943, Collection no. 13, Box 39, Dangberg Correspondence 1943, McCarran Papers.
17. Copy of a Western Union telegram from Patrick McCarran to John Dangberg, March 30, 1943; John Dangberg to Patrick McCarran, March 30, 1943, Collection no. 13, Box 39, Dangberg Correspondence 1943, McCarran Papers.
18. John Dangberg to the NRSOA members, April 15, 1943; John Dangberg to Arambel and Etcheverry, April 20, 1943; Pete Etcheverry to John Dangberg, April 19, 1943, Collection no. 13, Box 39, Dangberg Correspondence 1943, McCarran Papers.
19. Nash, *American West Transformed*, 50–54.
20. Grover B. Hill to Patrick McCarran, May 5, 1943, Collection no. 13, Box 39, Dangberg Correspondence 1943, McCarran Papers.
21. Patrick McCarran to John Dangberg, May 7, 1943; Earl G. Harrison to Patrick McCarran, May 15, 1943, Collection no. 13, Box 39, Dangberg Correspondence 1943, McCarran Papers.
22. John Dangberg to Patrick McCarran, May 27, 1943; D.A. Hughes to Patrick McCarran, May 26, 1943, Collection no. 13, Box 39, Dangberg Correspondence 1943, McCarran Papers.
23. Samuel P. Orth, "The Alien Contract Labor Law," *Political Science Quarterly* 22, no. 1 (March 1907): 51; Joseph Savoretti to Patrick McCarran, June 19, 1943; Patrick McCarran to John Dangberg, July 3, 1943, Collection no. 13, Box 39, Dangberg Correspondence 1943, McCarran Papers.
24. John Dangberg to Earl G. Harrison, May 28, 1943; John Dangberg to Tom Jolly, May 28, 1943; John Dangberg to D.A. Hughes, May 28, 1943; Patrick McCarran to John Dangberg, June 3, 1943, Collection no. 13, Box 39, Dangberg Correspondence 1943, McCarran Papers.
25. An affidavit submitted by John Dangberg and D.W. Park, Douglas County, State of Nevada, subscribed and sworn to before Notary Public, June 30, 1943, Collection no. 13, Box 39, Dangberg Correspondence 1943, McCarran Papers.
26. Ibid.
27. John Dangberg to Tom Jolly, June 30, 1943; John Dangberg to the Immigration and Naturalization Service, June 30, 1943; John Dangberg to Patrick McCarran, June 30, 1943; John Dangberg to Eva Adams, July 23, 1943, Collection no. 13, Box 39, Dangberg Correspondence 1943, McCarran Papers.
28. Patrick McCarran to John Dangberg, July 8, 1943; Philip G. Bruton to Patrick McCarran, July 15, 1943, Collection no. 13, Box 39, Dangberg Correspondence 1943, McCarran Papers.
29. Patrick McCarran to John Dangberg, July 26, 1943; John Dangberg to Patrick McCarran, July 30, 1943, Collection no. 13, Box 39, Dangberg Correspondence 1943, McCarran Papers.

30. John Dangberg to W.P. Wing, July 23, 1943, Collection no. 13, Box 39, Dangberg Correspondence 1943, McCarran Papers.
31. Kennedy, *Freedom from Fear*, 591–609; Summer Welles, *The Time for Decision* (New York: Harper & Brothers, 1944), 178–181; Robert Dallek, *Franklin D. Roosevelt and American Foreign Policy, 1932–1945* (New York: Oxford University Press, 1979), 262, 370–71; Joan María Thomàs, *Roosevelt and Franco during the Second World War: From the Spanish Civil War to Pearl Harbor* (New York: Palgrave Macmillan, 2008), 192–213.
32. *Statutes of the United States of America, Passed at the Second Session of the Sixty-Fourth Congress, 1916–1917, Part I: Public Acts and Resolutions*, 877; John Dangberg to Patrick McCarran, October 5, 1943; Earl G. Harrison to John Dangberg, October 18, 1943; John Dangberg to the Immigration and Naturalization Service in San Francisco, October 5, 1943, Collection no. 13, Box 39, Dangberg Correspondence 1943, McCarran Papers.
33. John Dangberg to T.B. Shoemaker, October 23, 1943, Collection no. 13, Box 39, Dangberg Correspondence 1943, McCarran Papers.
34. John Dangberg to the Immigration and Naturalization Service in San Francisco, October 5, 1943; John Dangberg to T.B. Shoemaker, October 23, 1943; John Dangberg to Patrick McCarran, October 5, 1943; Patrick McCarran to John Dangberg, October 8, 1943; John Dangberg to T.B. Shoemaker, October 23, 1943, Collection no. 13, Box 39, Dangberg Correspondence 1943, McCarran Papers.
35. T.B. Shoemaker to John Dangberg, November 4, 1943; John Dangberg to T.B. Shoemaker, November 15, 1943; Earl G. Harrison to John Dangberg, November 22, 1943, Collection no. 13, Box 39, Dangberg Correspondence 1943, McCarran Papers.
36. Earl G. Harrison to John Dangberg, November 22, 1943, Collection no. 13, Box 39, Dangberg Correspondence 1943, McCarran Papers; to learn more about the "Spanish Transatlantic Company," see Elena Hernández Sandoica, "La Compañía Trasatlántica Española: Una dimensión ultramarina del capitalismo español," *Historia contemporánea* 2 (1989): 119–20.
37. Earl G. Harrison to John Dangberg, November 22, 1943, Collection no. 13, Box 39, Dangberg Correspondence 1943, McCarran Papers; to learn more about the "Spanish Transatlantic Company," see Elena Hernández Sandoica, "La Compañía Trasatlántica Española: Una dimensión ultramarina del capitalismo español," *Historia contemporánea* 2 (1989): 119–20.
38. Enrique Moradiellos, *La España de Franco (1939–1975): Política y sociedad* (Madrid: Editorial Síntesis, 2003), 67–68.
39. John Dangberg and D.W. Park to the members of the NRSOA, December 4, 1943; H.K. Travers to John Dangberg, December 6, 1943, Collection no. 13, Box 39, Dangberg Correspondence 1943, McCarran Papers.
40. Patrick McCarran to John Dangberg, December 18, 1943; W.P. Wing to John Dangberg, December 21, 1943, Collection no. 13, Box 39, Dangberg Correspondence 1943, McCarran Papers.
41. John Dangberg to Gay Hill, December 7, 1943; John Dangberg to Patrick

McCarran, December 7, 1943; John Dangberg to Patrick McCarran, December 21, 1943; John Dangberg to Patrick McCarran, December 27, 1943, Collection no. 13, Box 39, Dangberg Correspondence 1943, McCarran Papers.

42. Axel Kreienbrink, "La Política de Emigración a través de la Historia del IEE," in *Historia del Instituto Español de Emigración: La política migratoria exterior de España y el IEE del Franquismo a la Transición*, Luís M. Calvo Salgado, María José Fernández Vicente, Axel Kreienbrink, Carlos Sanz Díaz, and Gloria Sanz Lafuente (Madrid: Editorial Síntesis, 2009), 15; María Encarna Nicolás Marín and María Alicia Alted Vigil, *Disidencias en el franquismo (1939–1975)* (Murcia: DM, 1999), 26–27; María Encarna Nicolás Marín, "Notas: La emigración española a América Latina," *Areas*, no. 7 (1986): 103.
43. Patrick McCarran to John Dangberg, December 28, 1943, Collection no. 13, Box 39, Dangberg Correspondence 1943, McCarran Papers.
44. Juan Pablo Fusi Aizpurúa, *España, 1808–1996: El desafío de la modernidad* (Madrid: Espasa, 1997), 181–184; Juan Francisco de Cárdenas to Patrick McCarran, December 29, 1943, Collection no. 13, Box 39, Dangberg Correspondence 1943, McCarran Papers.
45. Patrick McCarran to John Dangberg, January 3, 1944; John Dangberg to Patrick McCarran, January 14, 1944, Collection no. 13, Box 39, Dangberg Correspondence 1944, McCarran Papers; Moradiellos, *La España de Franco (1939–1975)*, 68.
46. John Dangberg to the NRSOA members, February 19, 1944; John Dangberg to James Petrini, January 20, 1944, Collection no. 13, Box 39, Dangberg Correspondence 1944, McCarran Papers.
47. E. Leuenberger to John Dangberg, February 21, 1944; Patrick McCarran to John Dangberg, March 1, 1944; Copy of a Western Union telegram from John Dangberg to Harry F. Hawley, March 6, 1944; Copy of a Western Union telegram from Alfonso Sario to Martin Goñi, March 8, 1944; H.K. Travers to John Dangberg, March 17, 1944; Copy of a Western Union telegram from John Dangberg to H.K. Travers, March 22, 1944; John Dangberg to Ralph Morrill, March 23, 1944; Application for a license to engage in a foreign exchange transaction, Nevada Range Sheep Owners Association, March 25, 1944; John Dangberg to H.K. Travers, March 27, 1944; John Dangberg to Patrick McCarran, March 28, 1944; some weeks later, on April 3, 1944, Cordell Hull sent a telegram to John Dangberg suggesting the same as Travers proposed before: "Make arrangements for deposit of transportation costs with New York Office of the steamship company concerned in order to expedite matter," Copy of a Western Union telegram from Cordell Hull to John Dangberg, April 3, 1944, Collection no. 13, Box 39, Dangberg Correspondence 1944, McCarran Papers.
48. Copy of a Western Union telegram from Martin Goñi to Alfonso Sario, March 23, 1944; John Dangberg to Eva Adams, March 23, 1944; John Dangberg to H.K. Travers, April 17, 1944; Copy of a Western Union telegram from Francisco de Amat to W.H. Wennhold, April 17, 1944; Copy of a Western Union telegram from W.H. Wennhold to Francisco de Amat, April 18, 1944; John Dangberg to H.E. Hosier, April 20, 1944; Invoice from Spanish Consulate paid to Farmers Bank of

Carson Valley, Inc., Minden, Nevada, Bank of America, April 20, 1944; Telegram cost, Farmers Bank of Carson Valley, Inc., Minden, Nevada, Bank of America, April 20, 1944; John Dangberg to Z.B. Jackson, April 24, 1944; Z.B. Jackson to John Dangberg, May 2, 1944; John Dangberg to T.B. Shoemaker, May 3, 1944, Collection no. 13, Box 39, Dangberg Correspondence 1944, McCarran Papers; "Nevadan After Herder," *Idaho Wool Growers Bulletin* 23, no. 9 (May 3, 1944): 7.

49. Copy of a Western Union telegram from Martin Goñi to John B. Dangberg, May 15, 1944; John Dangberg to District Director, Immigration and Naturalization Service in New Orleans, May 16, 1944; John Dangberg to H.K. Travers, May 17, 1944; John Dangberg to District Director, Immigration and Naturalization Service in New Orleans, May 22, 1944; Copy of a Western Union telegram from Patrick McCarran to John Dangberg, May 24, 1944; H.K. Travers to John Dangberg, May 24, 1944; Copy of a Western Union telegram from John Dangberg to Patrick McCarran, June 1, 1944; Copy of a Western Union telegram from Eva Adams to John Dangberg, June 2, 1944; John Dangberg to A.C. Ziegan, June 2, 1944; John Dangberg to Eva Adams, June 5, 1944; H.K. Travers to John Dangberg, June 7, 1944; Copy of a Western Union telegram from Eva Adams to John Dangberg, June 6, 1944; Copy of a Western Union telegram from Eva Adams to John Dangberg, June 7, 1944; A.C. Ziegan to John Dangberg, June 19, 1944; H.K. Travers to John Dangberg, June 23, 1944; John Dangberg to T.B. Shoemaker, June 27, 1944; Eva Adams to Grace M. Dangberg, June 27, 1944; John Dangberg to H.K. Travers, June 27, 1944; John Dangberg to Martin Goñi, June 30, 1944; John Dangberg to H.K. Travers, June 27, 1944; Copy of a Western Union telegram from John Dangberg to Patrick McCarran, July 10, 1944; Copy of a Western Union telegram from John Dangberg to H.K. Travers, July 11, 1944; Copy of a Western Union telegram from Patrick McCarran to John Dangberg, July 11, 1944; John Dangberg to Harmon E. Hosier, June 20, 1944, Collection no. 13, Box 39, Dangberg Correspondence 1944, McCarran Papers; "Spanish Basques Arrive in Nevada," *Reno Evening Gazette*, June 22, 1944, 16.
50. John Dangberg to H.B. Embach, October 12, 1944, Collection no. 13, Box 39, Dangberg Correspondence 1944, McCarran Papers; José Antonio Santesteban to John Dangberg, August 10, 1944, Box BSQAP 0114, 1944, Jon Bilbao Basque Library.
51. John Dangberg to W.P. Wing, August 8, 1944; John Dangberg to Frank Noriega, August 7, 1944; John Dangberg to Dan J. O'Doughlin, August 11, 1944; John Dangberg to Vernon Metcalf, August 11, 1944; H.B. Embach to John Dangberg, October 7, 1944; John Dangberg to James Petrini, October 10, 1944; John Dangberg to H.B. Embach, October 13, 1944; John Dangberg to James Petrini, November 9, 1944; John Dangberg to H.B. Embach, November 15, 1944; Copy of a Western Union telegram from M.C. Claar to John Dangberg, December 29, 1944, Collection no. 13, Box 39, Dangberg Correspondence 1944, McCarran Papers.
52. John Dangberg to Patrick McCarran, August 7, 1944; Patrick McCarran to John Dangberg, August 24, 1944; Cordell Hull to Patrick McCarran, September 7, 1944;

Eva Adams to John Dangberg, September 12, 1944, Collection no. 13, Box 39, Dangberg Correspondence 1944, McCarran Papers.

53. John Dangberg to M.C. Claar, January 1, 1945, Collection no. 13, Box 42, Dangberg Correspondence 1945, McCarran Papers; "Address, T.H. Gooding, President, to the Fifty-Second Annual Meeting of the Idaho Wool Growers Association," *Idaho Wool Growers Bulletin* 24, no. 1 (January 10, 1945): 2.
54. Eduardo María Danis to Alfred James Elliott, January 31, 1945, Collection no. 13, Box 42, Dangberg Correspondence 1945, McCarran Papers.
55. John Dangberg to W.P. Wing, February 15, 1945, Collection no. 13, Box 42, Dangberg Correspondence 1945, McCarran Papers.
56. John Dangberg to Eva Adams, February 12, 1945, Collection no. 13, Box 42, Dangberg Correspondence 1945, McCarran Papers.
57. John Dangberg to H.K. Travers, February 15, 1945; H.K. Travers to John Dangberg, February 25, 1945; John Dangberg to H.K. Travers, February 26, 1945; Eliot B. Coulter to John Dangberg, April 14, 1945, Collection no. 13, Box 42, Dangberg Correspondence 1945, McCarran Papers.
58. Kennedy, *Freedom from Fear*, 808–10; Pedro Antonio Martínez-Lillo, "La política exterior de España en el marco de la Guerra Fría: Del aislamiento limitado a la integración parcial en la sociedad internacional, 1945–1953," in *La Política Exterior de España en el Siglo XX*, eds. Javier Tusell, Juan Avilés, and Rosa Pardo (Madrid: Biblioteca Nueva, 2000), 325–30; Viñas, *Los pactos secretos de Franco con Estados Unidos*, 26.
59. John Dangberg to M.C. Claar, January 1, 1945, Collection no. 13, Box 42, Dangberg Correspondence 1945, McCarran Papers.

7

The Basque Immigrant Sheepherder Question and U.S.–Spanish Relations during the early Cold War, 1945–1954

On April 30, 1952, Rafael Fagoaga, a twenty-three-year-old Basque immigrant living in Nevada, declared for the first time his intention to become a citizen of the United States. Fagoaga was born on November 7, 1928 in the town of Lesaka in northern Navarre. Like other members of his family and many other youths of Lesaka, he had immigrated to the United States to become a sheepherder in the Far West. Fagoaga had left Spain during the early years of Francisco Franco's dictatorship in the turbulent context of the political repression and economic depression of the late 1940s. When the opportunity to immigrate and work as a sheepherder in the Great Basin presented itself to him, Fagoaga did not think twice before leaving for America.

On February 27, 1948, at nineteen years of age, Fagoaga flew by commercial airliner, traveling from Madrid to Reno, entering the country legally via New York's airport. It was his first time on a plane and being so far away from home. Soon after his arrival in Reno, Fagoaga traveled on to northeastern Nevada, settling in Elko County where he worked as a sheepherder. Evidently not wanting to return to Spain, after four years in the United States, Fagoaga enthusiastically applied for U.S. citizenship in the Fourth Judicial District at the courthouse of Elko County. He filled out the personal physical description on his citizenship application as follows: "Sex, *male*; color, *white*; complexion, *dark*; color of eyes, *brown*; color of hair, *brown*; height, *5 feet 6 inches*; weight, *165 pounds*; visible distinctive marks, *none*; race, *white*." As soon as he received his papers, Fagoaga quit the sheepherding job and began working as a bartender at the Star Hotel in the town of Elko where he met his future wife—Jean Pilar Martinez, a second-generation Basque-American girl who worked as a waitress at the same boarding-house. They eventually married and raised together a family of three children.[1]

When considering the history of Basque immigrant sheepherders in America, the case of Rafael Fagoaga (1928–1996) is significant in many ways. The short length of his career in the sheep industry as a wage laborer contradicts the common narrative of Basques as an ethnic group destined only for sheepherding work. Fagoaga was one of the first Basque immigrants who were formally hired as a sheepherder under the labor-contract terms established by the Franco regime in the late 1940s. Moreover, he was one of those early Basques who came to the United States by plane, instead of by ship and then by rail to the West. He did, of course, have to reimburse his employer for his transportation, something that could be quite expensive at that time.

In the late 1940s and early 1950s, like almost every aspect of life in the United States, Basque immigration was conditioned by the postwar international context. In the aftermath of World War II, as long as the quota system remained in force, the problem of bringing greater numbers of Basque immigrants persisted. As the Cold War intensified, relations between the United States and Spain warmed. The Franco regime saw the Basque immigrant sheepherder question as a means to improve relations with the United States even further. Franco's regime soon understood that it could benefit by opening its doors to allow Basques to emigrate to the United States. By the second half of the 1940s, Franco's regime had begun to take a direct interest in the Basque immigration to the American West. Although the U.S. government waited some years to formalize diplomatic relations with Franco's Spain, trade increased along with more open and lenient policies on Basque immigrant labor from both countries. The geopolitical factors cannot be ignored in this development.

The present chapter sets the Basque immigrant labor issue in the context of diplomatic negotiations between the United States and Spain that eventually crystallized into the Pact of Madrid in 1953—the military-related cooperation contract between the two governments that officially ended the ostracism against Spain from western liberal democracies. This chapter analyzes how the process of importing more Basque labor from Spain was closely intertwined with the evolution of U.S.–Spanish relations during the early Cold War. It explores the interests of both Patrick McCarran and Francisco Franco in encouraging further Basque immigration to the American West.

The Struggle for Permanent Status, 1945–1947

In the aftermath of World War II, in the eyes of American ranchers and the general public, Basque immigrants were increasingly viewed as a highly valued "race" in the American West. The political efforts to persuade lawmakers to allow increasing numbers of Basque immigrants during wartime reinforced the image of the Basques as "good sheepherders." Furthermore, public opinion considered Basques a distinct group of people who had become different through long isolation. This sense of uniqueness and rarity made Basques seem different from others and increasingly desirable as workers, especially as sheepherders. As Roy W. McNeal, the Hermiston, Oregon, irrigation district supervisor, said about the Basque immigrant community settled in Jordan Valley in 1945: "The Baskoes [*sic*] are a fine people—extremely patriotic to the United States, hard workers and law-abiding."[2]

While Basques had a reputation of being admirable workers, the sheep industry, where that reputation had been built, was facing a deteriorating economic situation. After World War II, although the agricultural economy continued to grow, the economy in Nevada and other Western states changed in remarkable ways that significantly altered the sheep industry. Besides increasing federal regulation of livestock grazing on public lands and dietary diversification, the American sheep market was further affected by the importation of woolen goods from other countries. This foreign competition caused prices to fall in the sheep industry. While ranchers continued to graze livestock on Nevada ranges, they would do so largely on the public domain under the supervision of grazing districts administered by the new Bureau of Land Management (BLM) created in 1946 by executive order to replace the Grazing Service. Despite confronting new regulations in grazing districts, foreign competition, and new synthetic fibers, sheep ranching continued as did the demand for immigrant workers to replace older sheepherders as well as those who had been drawn into urban occupations in the New West.[3]

In late 1945, Nevada ranchers continued to face urgent problems concerning the importation of Basques and the legalization of status for those immigrants whose work permits were soon expiring. The

NRSOA persisted not only in its efforts to bring greater numbers of Basque workers from Spain to the Great Basin, but it also started working to find ways to keep Basques in the country. On September 29, 1945, John Dangberg and D.W. Park, on behalf of the western Nevada woolgrowers, submitted an application at the office of the Immigration and Naturalization Service in Reno for the extension of visas for fifty-nine sheepherders of the seventy-four Basque immigrants who had arrived a year before. Dangberg and Park raised the previous arguments to justify the extension of the Basques' temporary stay. According to them, the equation was as follows: "If the sheep industry is to be preserved in the particular sections skilled men are needed and these men are the only ones available."[4] Dangberg and Park wrote that ranchers had so far expressed their satisfaction with these Basques and the immigrants themselves desired to remain in that occupation as well. In the case that these men were required to leave the country before October 15, 1946—the date of expiration of their visas—they said, the Nevada sheep industry was going to suffer an important economic loss:

> That the circumstances requiring the continued presence in the United States are that the services of these men are required where they are located to preserve and to continue the sheep industry. That the industry would suffer disastrous effects if the extension was not granted in that it would mean the reduction of herds by from 75,000 to 100,000 ewe sheep. That the employment of the men is continuous and will remain so during the life of this extension.[5]

Along with this declaration was attached a list with the names of the fifty-nine sheepherders and their employers, as well as some other legal information pertaining to their immigration status.[6]

The NRSOA had made the application through the Immigration and Naturalization Service for an extension of the duration of the visas of the fifty-nine sheepherders for one year—from October 16, 1946, to October 15, 1947. This application, however, was denied. The Immigration and Naturalization Service had determined that there was already sufficient labor of this type in the United States and therefore, the competing foreign labor must depart. However, Nevada ranchers demanded only Basques. Although this application was not immediately granted, John Dangberg continued working to renew all the legal documentation of these Basque aliens with both the American and Spanish immigration authorities. Dangberg contacted the Spanish consulate in San Francisco to have the Basque sheepherders' passports renewed and ready for the time the immigration authorities granted such permission. Dangberg never threw in the towel in his project to enable these fifty-nine workers to remain in the country and persisted in his efforts to secure Basque labor in the rangelands of Nevada.[7]

In addition to the immigration issues concerning visa renewals, the postwar period revealed many cases of Basque immigrants who were residing illegally in the United States. During the Second World War, many Basques had entered the United States illegally or legally, either with temporary work permits or on tourist visas, and then stayed there after the expiration of their permits. In the early postwar period, these cases were sanctioned according to the law at the time when other non-Basque immigrants were also in the same situation. By then, the U.S. government began considering measures to provide legal status to those aliens existing in the United States.

On August 7, 1946, the Seventy-Ninth Congress passed a federal private law "for the relief of certain Basque aliens." This joint resolution empowered and authorized the attorney general of the United States to cancel deportation proceedings in the particular cases of numerous Basque immigrants who were illegally residing in Utah and Eastern Nevada. These Basque immigrants entered the United States during World War II and were admitted as "seamen," although eventually ended up working as sheepherders. After the expiration of their work permits, those Basques remained illegally working as sheepherders. These "aliens," the private law 818 read, "shall be considered as having been admitted for permanent entry as of the date of their actual entry on the payment of the visa fees of $10 and head taxes of $8 per person." Under this act, the Secretary of State James F. Byrnes "shall instruct the proper quota-control officer to deduct nine numbers from the Spanish quota for the first year that the said Spanish quota is available."[8]

Like in Nevada, the extension of temporary stay visas for those Basque sheepherders in the other western public-land states was also considered an urgent issue. In general terms, because the availability of manpower had increased after the war, the Immigration and Naturalization Service was less willing to allow Basque immigration or extend permits of those who were already residing in the country, because "labor of the type for which the extension is requested is available in the United States." Thus, when many sheep ranchers in the West tried to extend the working permits of their Basque workers, the Immigration and Naturalization Service first denied them on the grounds that there was already availability of this kind of labor for the postwar market. For example, Arthur H. Bohoskey, president of the Yakima Sheep Company in the state of Washington, had applied for an extension of visas to March 19, 1947, for seven Basque immigrants under his employ. But in November 1946, the Immigration and Naturalization Service's central office in Philadelphia refused the application.[9] In those cases where the ranchers could not legalize their contracted workers' residence status in the United States, these Basque immigrant workers had to return to Spain.

Western woolgrowers continued to respond to the Basque labor problem by joining together in their fight to obtain extended visas for the sheepherders. On November 20, A. E. Lawson, secretary of the Washington Wool Growers' Association, wrote to John Dangberg: "I wish to advise you so that you may be on your guard to protect yourself in this matter, and also forestall any effort by the Immigration authorities to send all of these Basque herders back to Spain." Then, John Dangberg contacted other western woolgrower associations suggesting a mobilization of all the sheep ranchers and together to report the scarcity of "skilled range sheepherders" in order to justify to the immigration authorities the necessity of keeping the Basque aliens in the country and importing more from Spain. Eventually, the continuous pressures from several associations and their respective representatives to the federal government pushed the Immigration and Naturalization Service to lessen its restrictive outlook.[10]

On December 16, 1946, Joseph Savoretti, assistant commissioner of the Immigration and Naturalization Service in Philadelphia, informed J. M. Jones, secretary and treasurer of the National Wool Growers Association, that in view of the apparent shortage of "experienced and responsible herders . . . no action will be taken toward requiring the departure of the above referred to importees for sixty days in order to afford the Yakima Sheep Company an opportunity to present evidence showing that the Emergency Farm Labor Office at Spokane was unable to furnish the labor desired."[11]

Then, the National Wool Growers Association took the matter of "the lack of dependable sheepherders" in the American West to the

Department of Justice, Immigration and Naturalization Service. The problem of Basque labor plagued the sheep industry and concern spread to other areas in the American West. On January 12, 1947, the fifty-fourth annual convention of the Idaho Wool Growers Association, president T.H. Gooding, in his opening address, said: "Our labor situation is still a vital problem. We are still unable to obtain any new men from Spain. The Department of Immigration has turned down all requests for importation of Basques. Under the law, new labor can only be imported where there is an extreme shortage."[12]

Dangberg, for his part, once again requested help from the Senator Pat McCarran on this issue. Both continued working together on this issue.[13] In the second half of the 1940s, McCarran stood fast in the defense of the interests of his Nevada constituents. By then, McCarran had become, as historian Jerome Edwards has said, "the undisputed political boss" in Nevada. Further, while he rose to national prominence as a senator in the Cold War political era, McCarran increasingly paid more attention to international issues, as he became more convinced that global economic forces and other foreign factors affected directly or indirectly the welfare of Nevada. Once he had achieved this influence, McCarran was in the position to embark on other political adventures of international character concerning issues related to trade, geopolitics, immigration, and the labor market. In this moment of his political career, the senator from Nevada would fight on every front to bring more Basque immigrant workers from Spain.[14]

As a result of the efforts of McCarran on behalf of the sheep industry in Nevada, on March 14, 1947, eventually, Joseph Savoretti informed Dangberg that the Immigration and Naturalization Service had granted an extension to stay for the fifty-nine Basque aliens until March 14, 1948; "upon consent of surety, provided that these aliens are employed only as sheepherders." Thus, the Immigration and Naturalization Service granted an extension of residence for these Basque immigrants on condition that they had to post a new bond for maintenance of said workers' status and guaranteeing their departure from the United States at the expiration of their temporary admission. In early 1947, Dangberg proceeded to make the necessary arrangements to get everything done immediately. This included the renewal of the Basque sheepherders' passports and Certificates of Nationality through the Spanish consulate in San Francisco, which facilitated most of this bureaucratic process. Although the granted extension permits for an additional year solved the problem temporarily, Dangberg considered it insufficient. He was convinced that solving the Basque labor problem was more than just regularizing the legal status of those Basques residing in the United States.[15]

In spring 1947, Nevada representatives began taking action on the Basque labor problem. On April 15, the State Farm Labor Advisory Committee of Nevada adopted a resolution declaring their intention to extend the Emergency Farm Labor Program to December 31, 1947 (the labor plan had been established in 1942 to solve the generalized agricultural labor shortage), "including the importation of foreign workers as our experiences indicate that the domestic workers are still too unsettled and the supplies are not yet sufficient to fill all the needs for Agricultural workers." Thus, for the year 1947, the Farm Labor Advisory Committee of Nevada considered it necessary to continue depending on immigrants "on certain types of work for which domestic workers cannot be secured." After December 31, 1947, this committee recommended development of a longtime agricultural program under the Federal Extension Service of the Department of Agriculture in order to create a new agricultural-trained workforce. Meanwhile, Nevada ranchers continued working to set up their Basque workers' immigration papers. The Immigration and Naturalization Service for its part kept watching closely the legal status of the Basque immigrants as contract laborers admitted under the waiver of the Alien Contract Labor Law.[16]

In early October 1947, John Dangberg sent a letter to all the NRSOA members requesting information from them concerning the sheepherders who had entered the country in the summer of 1944 and whose permission of stay expired on March 14, 1948.[17] The NRSOA sent this letter to their members with a view toward compiling data to use it "either to secure extensions of stay or to offer data supporting a bill to give all the bona fide herders permanent status in the country." The urgency among sheep operators was clearly evident. Before the postman had even reached his mailbox, on October 8, Reginald Meaker, owner of the Maud Wheeler Sheep Company, had written a letter to the Dangbergs saying that he was hoping to extend the working permit of one of his Basque herders.[18]

Ranchers continued praising highly the Basque workers and expressed satisfaction with their employment. All of them agreed that they wanted these Basque immigrants in the country working for them and the herders themselves had declared their desire to stay. It should be noted here that a great part of the sheep ranchers of the NRSOA were Basques themselves who had emigrated in the early part of the twentieth century and had acquired prominent economic positions. This gave them the capacity to act on behalf of their compatriots.[19]

Some of the Basque sheepherders brought by the NRSOA in the summer of 1944 were employed by sheep ranch owners outside of Nevada in Idaho and Washington. Dangberg offered his assistance to the Idaho Wool Growers Association to help them see how their firms should proceed in an effort to extend the working permits of their laborers.[20]

Dangberg gathered the letters from the Nevada ranchers as a proof not only to extend these sheepherders' stay, but also to use this data as a means to do something effective about getting them into the country permanently. On December 10, the NRSOA submitted an application to the Immigration and Naturalization Service in Philadelphia to extend the stay of forty-eight Basque sheepherders. This was a part of a temporary answer to a broader problem, as all the western woolgrower associations believed that the seeds of a lasting solution lay in the establishment of a specific plan to import Basques that would answer to the demands of the associated sheep ranchers. In October 1947, from Twin Falls, J. H. Breckenridge, the president of the Idaho Wool Growers Association, had said: "Unless there is a change in the minds of young Americans concerning working with sheep, if our industry is to continue, I would say that not only will these work contract Basques have to stay but more young men from their country must be imported." In December 1947, McCarran had already taken the first steps to put the forty-eight Basque sheepherders on permanent basis in the United States and began considering new formulas for the solution of the Basque sheepherder question.[21]

Between McCarran and Franco: The Basque Sheepherder Question, 1947–1950

Once the war was over, Francisco Franco's Spain was being excluded from the reconstruction process orchestrated primarily by the United States. First in San Francisco and later in Potsdam, the United States reaffirmed its position to isolate and exclude Franco's regime from Marshall Plan aid. In the Cold War's first years, the Spanish regime could do nothing but remain silent and wait. Other issues soon absorbed the international community with the worldwide struggle between the United States and the Soviet Union being the most prominent. In the new international context, the Franco's regime could exploit its strong anticommunist identity to its advantage and thus gain favor with the United States. As the Cold War worsened, the regime correctly assumed that the United States would seek to restore diplomatic relations in its mission to contain the expansion of the Soviet Union.

In the process, the international community pressured Franco to implement liberal democratic reforms. Nevertheless, despite continued pressure from the Western world, Franco's regime strongly resisted and took no steps at all toward real political democratization in Spain. Instead, the Spanish dictatorship pursued a wait-and-see foreign policy toward the United States. The strategy held consequences for Spain, which, in the absence of American financial aid, endured a long postwar economic depression. The Truman administration, for its part, appeared in no hurry to move. President Truman despised Franco (and McCarran) and before developing a new policy with the Spanish dictatorship, a change in presidential administrations was necessary in Washington. Nevertheless, it simultaneously promoted mutual commercial relations and negotiated to solve issues involving the two countries. This included the case of the recruitment of Basque workers for sheepherding in Nevada.[22]

The postwar anticommunist movement in the United States looked upon Franco's regime with some approval and certainly did not see it as a threat to American national security. There was in some of the political climate of the early postwar era agitation against the reforms of the New Deal and criticism of Truman's foreign policies. Republican critics charged the Truman administration with being soft on communism. In the emerging East–West struggle, some conservative voices pushed for recognition of Franco's regime to bring Spain into the American orbit. The result of these Cold War circumstances was closer relations between the United States and Spain. Patrick McCarran, who had familiarized himself with the political situation in Spain through his efforts to bring Basque immigrant workers to Nevada, was a leading proponent of closer relations. Soon McCarran came to admire and esteem the Spanish dictator.[23]

The Cold War formed the background for negotiations over Basque immigrant workers from Spain. In the postwar period, McCarran, now in his seventies, began to turn his attention to international issues. During the late 1940s, McCarran intruded himself into negotiations between the United States and Franco's Spain. He also emerged at the forefront of the post-World War II "Red Scare" along with Joseph McCarthy of Wisconsin. Both were some of the first public voices in support of Franco's dictatorship. For McCarran the issue of Basque sheepherders played an important role in shaping his attitude toward Franco's Spain.[24]

On March 12, 1947, President Truman asked Congress to provide economic assistance to Greece and Turkey in the containment move against the Soviet Union. With the Truman Doctrine, the United States assumed a leading role in the worldwide struggle against the Soviet Union. Francisco Franco welcomed the news. The Spanish regime swiftly reaffirmed its anticommunist credentials and increased its anticommunist rhetoric against the Soviet Union as part of the strategy to receive economic aid from the United States. The international context opened many doors for Franco's Spain. By late 1947, in the political circles of Washington, D.C., representatives of both parties and other prominent figures in the government started to view Franco's Spain in a more favorable light. It became a bone of contention between the administration and the Republican Congress around which the "Spanish lobby" formed.[25]

In 1948, as historian Ángel Viñas has noted, Washington, D.C. had become the most important foreign capital for Franco's regime. A year before, Spain stood outside the European Recovery Program, known as the Marshall Plan. Only improved relations with the United States could assure Francisco Franco financial assistance from the United States. Franco sent José Félix de Lequerica y Erquiza to the U.S. capital for the purpose of building bridges in the hopes of gaining economic aid in the future. In order to undertake the mission, Lequerica was appointed "Inspector of Embassies, Legations, and Consulates" by Franco.[26]

Once settled in Washington, Lequerica led an effective interest group—the "Spanish lobby"—constituted on behalf of Franco's regime, which lobbied Congress to win recognition and economic aid for Spain. In this diplomatic venture, Lequerica relied upon Germán Baraibar and Eduardo Propper de Callejón, who were in charge of the business and commercial mediation issues. They were ardently Catholic, fervent anticommunists, military planners, members of the Republican Party, and businessmen who were part of the Spanish lobby. As Viñas has said, the Spanish lobby represented a diverse interest group of different religious, economic, and political backgrounds. They often disagreed among themselves about many issues, but agreed on the necessity of supporting Franco's regime as a strategy in the Cold War.[27]

Patrick McCarran became an important factor in the conservative effort to build a relationship with Franco's Spain. He had many things in common with the members of the Spanish lobby. He was a Catholic; although a Democrat, he opposed the left-leaning New Deal; he was a dedicated anticommunist who valued Spain's strong opposition to any left-wing organizations, as well as its geopolitical location in the Cold War strategy. Also, he acted on behalf of the commercial interests of his constituency of Nevada: the sheep ranchers who were anxious to bring Basque herders from Spain to their stock operations. While McCarran supported their efforts, on the other hand he opposed opening the doors to other immigrants, especially war refugees. In 1948, Congress passed the Displaced Persons Act authorizing the admission of about two hundred thousand refugees of World War II into the United States. Like the Spanish question, the administration of the displaced persons program created division and tension within Congress and in President Truman's cabinet. McCarran opposed the law and worked persistently to prevent Jewish refugees or what he considered undesirable aliens from entering the United States, while he welcomed others (the Basques) whom he considered desirable.

By early 1948, Pat McCarran held two positions in two prominent investigating subcommittees in the Senate, offering him a certain level of power to decide on immigration issues. He chaired a subcommittee of the Senate Judiciary Committee, which conducted internal investigations of U.S. immigration policy, and he was appointed a member of a standing subcommittee of investigations of the Immigration and Naturalization Service. This was excellent news for the Nevada sheepmen who were anxious to settle the forty-eight sheepherders in the country and recruit increasing numbers of Basques from Spain. John Dangberg placed a great deal of confidence in the Nevada senator. On January, 10, 1948, Dangberg wrote McCarran: "Now that you are on a committee to study immigration we think there will be some constructive thinking done on this whole business."[28]

In early January 1948, McCarran began drafting a bill to grant permanent admission to forty-eight Basque sheepherders already in the country. On January 12, McCarran introduced in Congress Senate Bill S. 1973 "for the relief of certain Basque aliens," which aimed to allow the forty-eight Basque sheep workers to work and reside in the United States and make it so the Nevada sheepmen were not required every year to apply for an extension of working permits.[29] The bill went to the subcommittee of the Senate Judiciary Committee of which McCarran himself was a member. On the same day on which McCarran introduced the bill, the senator from Nevada wrote to Dangberg saying that, "it will be impossible to secure action until after a formal report has been requested from and submitted by the Immigration and Naturalization Service. I shall do what I can to see that this report is expedited." Dangberg was pleased with McCarran's letter. Immediately after, Dangberg contacted other secretaries of western woolgrower associations to inform them about the new bill and encourage them to contact their respective congressional delegations in order to find similar solutions to the Basque labor problem.[30]

Before the immigration authorities could make a report on S. 1973, federal law required an investigation of all the forty-eight Basque sheepherders by conducting interviews with them in the local field offices. Dangberg made all the necessary preparations to undertake the local interviews. The Immigration and Naturalization Service required interpreters because the Basque workers did not know English. Eventually, the forty-eight Basques were permitted to stay. However, there was still a pressing need for more sheepherders. In March 1948, in the fifty-fifth annual convention held in Boise, the Idaho Wool Growers Association complained that: "it has been almost impossible to import alien Basque sheep workers into this country. It takes miles of political red tape and fifty-fifty chance of getting competent men." In the spring of 1948, as a new lambing season approached, sheep ranchers demanded their associations push for more Basque labor.[31]

Karl V. King and the Three-Hundred Basques, 1948

On another front the attorney Karl Vernon King of Utah earned a reputation for defending the rights of Basque immigrants. Since the early 1930s, King had represented many Basque aliens before the immigration judges in deportation proceedings. As the labor shortage worsened in the late 1930s and early 1940s, sheep ranchers, who desired to bring in Basques, began contracting King to handle legal issues related to their admission to the country.[32] As a result of his experience in getting necessary extensions for Basque aliens, King soon acquired a reputation as the main lawyer and champion of the Basques in the American West. Later on, King himself recognized that through his "association with these Basque sheepherders, I have come in contact with many sheep owners and operators."[33]

By 1947, King had developed considerable expertise on the question of Basque immigrant labor and on extending the stay of aliens in the United States. In January 1947, the ENSGA and operators in southern Idaho individually contracted King to apply for increased numbers of Basque immigrants from both France and Spain. After he had filed the original petition, the Immigration and Naturalization Service and the Agricultural Extension Service held hearings to determine the economic necessity of the importation of Basque immigrants. King himself attended the hearings to follow the investigation closely.[34]

On September 15, 1947, after the investigation had concluded, the Immigration and Naturalization Service granted a permit to import 158 Basques into the United States. By October, as the word spread quickly in the provinces of Biscay and Navarre about the opportunities in the West, 300 Basques applied to the police for passports. Most of them were advised by their relatives who were already living in the United States to apply. The large number of requests for passports coming all at one time disturbed the Franco regime in Madrid to the point where it stopped the issuing of all passports.[35]

The Spanish regime for its part became directly involved in this Basque labor recruitment process, motivated by the desire to get the financial support from the United States. The Franco regime's strategic shift from an anti-Western position to a more cautious approach based on grounds of convenience also reflected its stance regarding the recruiting of Basque immigrant workers. The central government asked the Spanish embassy in Washington, D.C. to undertake an investigation on this question. In November 1947, the Spanish Embassy requested King fly to Washington and discuss in person the question of the Basque sheepherders with the Spanish officials. At first, the embassy had told King that they would not issue more passports for Basques unless western woolgrower associations put pressure on the U.S. State Department and on the Senate for a labor treaty. However, the possibility of arranging a labor program similar to the 1942 Bracero program would not guarantee Basque immigrants upon their entry to the United States any permanent status. Two weeks after the meeting, King persuaded the Spanish Embassy that a labor treaty of this kind with the United States was impossible and finally they did agree to sign a labor contract.[36]

After those meetings, King left Washington feeling very positive on the issue of Basque immigrant labor. He returned to Salt Lake City and on his arrival there, he found a letter from the Spanish Embassy advising that the Francoist government required woolgrowers to pay a "Letter of Call Tax" (*Impuesto de Cartas de Llamada*), to be paid at the Spanish Consulate in San Francisco. King with the assistance of John Landa, a Basque sheepman, began working with the Spanish consul to expedite the process and pay these taxes to Franco's regime. While having some complications in the procedure, King and Landa, on behalf of the western woolgrowers, redrafted the labor contract eight times, provided a property bond in the amount of $50,000, and paid both the existing visa tax amounting to $5.50 and the "Letter of Call Tax" in the amount of $3.96 for each man. Also, King proceeded to arrange all the details for the trip through Trans World Airlines with headquarters in New York. Basque immigrants would no longer embark on the classic steamship travel crossing the Atlantic Ocean, but fly more comfortably on a plane. In late 1947, after two weeks of intense negotiations with the Spanish officials, making calls and doing paperwork, King and Landa fulfilled everything required and the Spanish Consul issued the *Cartas de Llamadas.* Just like the other previous times, the immigrants' sponsors paid all the expenses and services in advance. With the added taxes by Franco's regime, sheepmen had to pay in total about $1,100 per man.

Everything needed to get done before February 15, 1948—expiration of the deadline fixed by the Immigration and Naturalization Service.[37] Nonetheless, despite the urgent timeline, some administrative problems in processing the sheepherding applicants through travel agencies in Bilbao and Pamplona delayed the procedure, and because of the red tape involved in the process, it was impossible to get the laborers to the United States within the time allotted. King reported the new situation about the group of Basques to the Immigration and Naturalization Service. He convinced the immigration authorities to extend the deadline to March 15 and let him travel to Spain to deal with the problem personally. King got permission and traveled to the Basque Country. Once there, he saw that the aspirant Basque "sheepherders were completely unfamiliar with the procedure necessary to obtain a Spanish passport." Thus, these Basques had to find help among friends and from travel agencies to obtain their passports. Typically, these agencies and even some friends abused the workers by charging exorbitant fees for obtaining full legal documentation. Moreover, King explained that the situation tended to differ with the local authorities in Bilbao and Pamplona. He described the situation like this:

> In Bilbao we generally got along very well with the police. All matters worked smoothly in Pamplona and there we did business with a very fine travel agency who, as far as I have yet heard, treated the herders very fairly, and there was no fee gouging; however, I did have difficulty in Pamplona with the police. The police insisted that they had only one stenographer who could get out passports and that stenographer could turn out only one passport a day and there was nothing they could do about it. We finally asked them if they could employ additional stenographers at our expense and get the passports out a little faster. The next day they advised us that they could have all passports out in three days if we would pay 25,000 Pesetas, or approximately $1,500. After considerable deliberation as to the advisability of paying such a sum, I concluded that I had better pay it or abandon hopes of obtaining any men from Navarre.[38]

After they paid, Spanish officials issued immediately the passports for the boys from Navarre who arrived ahead of those from Biscay. Because the group of Basques was so large, Franco's regime considered King as an exporter of a commodity operating in Spain. It required him to pay for an "export license" in the amount of $1,500.[39]

After three weeks in the Basque Country scrambling to solve the problems of passports with the Spanish administration, King made his way back to New York. By that time, the issue of the displaced persons was dominating the American political debate and profoundly shaping the politics of immigration in both houses. While some in Congress opposed opening the gates to the Jewish refugees, they persistently worked to open them to the Basques. Several displaced persons' groups and labor unions strongly complained about the priority given to Basques by the

Immigration and Naturalization Service. Furthermore, this issue raised criticism from some Democratic representatives and liberal organizations. Responding to an emerging liberal criticism of the late forties, the Immigration and Naturalization Service adopted a more restrictive policy stance on the admission of the Basques, all of which made more difficult the employment of these immigrants. Now, the Immigration officials demanded the western woolgrowers obtain certification from the Employment Service as a tangible proof of the labor situation in the sheep industry before any more Basque men could obtain legal permission to enter the United States. To the woolgrowers' frustration, the Employment Service went on to determine that the supply of labor of this kind was sufficient and that there was no real need for importing increasing numbers of Basque immigrants.[40]

The National Wool Growers Association complained that the federal government and in particular the Employment Service was prioritizing American labor that was, in their point of view, inexperienced and not interested in taking these jobs. J. M. Jones, secretary and treasurer of the National Wool Growers Association, said that the main objective was always securing "men who will stay with the herds and not have to come to town every three or four days or every week." He explained as follows how the federal government's policy sponsoring American national labor was detrimental for the sheep industry: "It seems highly essential that we acquaint the U.S. Employment Service with the requirements of a sheep herder. Some of their offices do not seem to appreciate the fact that a surplus of farm labor does not mean that there is a surplus of qualified, experienced, dependable sheep herders."[41]

Furthermore, Jones blamed the federal government for not allowing the recruitment of Basque immigrants who were the only ones willing to accept those arduous and tedious working conditions in the range sheep industry. The Truman administration, he contended, was "throwing everything they can in the way to prevent this Basque importation." Jones suggested importing Basques via the displaced persons provisions: "As most of you know, the United States is going to permit large numbers of displaced persons to enter the United States and it would appear that we are entitled to get our share of men who could be of use to us within that group."[42]

Right after Jones wrote these words, King telephoned from New York reporting that all Basque immigration was stopped and approximately three hundred Basques already in the country were under a deportation order. King also got in touch with M.C. Claar, W.P. Wing, John K. Robbins, and several U.S. Employment offices in the American West. The immigration authorities denied their entry based on the assumption that there were already enough workers for such jobs in the United States. On March 15, 1948, the Immigration and Naturalization Service had ruled in this way against the importation of Basque labor based on the reports of the Employment Service, which affirmed that there was already a surplus of labor in the United States. The central Employment Service had received information pertaining to the sheep labor situation all over the West from its field offices of those states. Typically, the western woolgrower associations had contacted their respective offices in order to report the labor shortage and continuously demanded immediate action to solve the problem. In this particular case, the problem concerned the report sent from California.[43] The local employment offices of Bakersfield, Fresno, Sacramento, Marysville, and a few other areas in California reported to the central office that there was no need to import increasing number of Basque immigrants in California. Apparently, the employment officials in California could not certify a shortage, because many Basque ranchers did not place requests for increasing herders through the local federal offices. What grabbed the attention of the immigration authorities at Ellis Island was that more than 50 percent of the total group of immigrants was set to go to California (160 of 300 Basques). To the immigration authorities, these numbers did not match with the report from the Employment Service officials, which revealed no labor shortage. Offices in Boise and Denver also sent "bad" reports. In response, the Immigration and Naturalization Service detained this group of 300 Basques and they had to wait at Ellis Island until clearance was issued on April 23, 1948.[44]

Responding to their grazing constituents, Senator McCarran, Congressman Charles H. Russell from Nevada, and Congressman Hal Holmes from the state of Washington made efforts to get the Immigration and Naturalization Service to delay action on deportations until they received a further report from the Employment Service. John Dangberg got upset "in view of our present move to have our group of men admitted for permanent residence."[45] In the end, this small mistake throws it all out of whack.

As could be expected, western woolgrowers' reactions were immediate. On March 19, 1948, W.P. Wing from California criticized the

Employment Service for failing to accurately report the labor situation in the sheep industry. Wing explained the labor situation as follows:

> There is a vast difference between an ordinary farm worker and a sheep herder. Farm labor in most instances is under constant supervision of the owner or foreman of the farm. The worker knows from day to day exactly what to do. In the case of a sheep herder, he is out on the range or in rough mountainous terrain for many days without supervision. He has to know how to handle a herd to avoid losses and in many instances he is handling an investment worth at least $50,000. It takes several years to properly train men for this work.[46]

Wing asserted that the responsibilities of sheepherders in the Interior West were enormous since they were in charge of nearly all the inputs into sheep production. To Wing, "Basque herders are experienced in herding in this type of country" and contradicted the Employment Service report, saying that "our domestic labor supply does not care to work and will not work under difficult conditions of many of these rough areas." Basques seemed to be among those few immigrant groups who accepted this job of long hours and very poor working conditions.[47]

Because the situation was urgent, King flew to Washington, D.C. to lobby for obtaining permission to allow these Basque immigrants to enter the country. He first asked for help from the western congressional delegation. He took the matter up with the Employment Service in Washington. Also, King met with: Walter Erb, acting assistant director for Farm Placement; David M. Fessenden, who was directly in charge of making the reports of availability and economic necessity to the Immigration and Naturalization Service; and W. S. Dutton of the Forest Service. According to King, Dutton was "extremely friendly and cooperative." Dutton had stated to the Employment Service officials that their experience with native-born Americans in the national forests was discouraging and "that the Basques were dependable and reliable, and handled the sheep on the forests in a very satisfactory manner and in the best interest of the public." But in general, King found that there was "a little resentment" among different federal officials. King considered it of vital importance to make "their wants, needs, and requirements known to the U.S. Employment Service."[48]

Following the advice of King, western sheepmen got to work with the assistance of their representatives. On April 23, Pat McCarran informed

Dangberg that he was actively working on the Basque sheepherder question. McCarran proclaimed that "Basques should be granted special consideration only because of the urgent need for sheepherders, and that if they choose to leave that occupation, they are no more deserving of special consideration than any other class of aliens." In mid-May, Dangberg and D.W. Park submitted an affidavit similar to the previous ones indicating the urgent need of importing Basque immigrant labor. Dangberg and Park wrote: "Affiants each for himself further state that the range sheep industry of the states of Nevada, California, Washington and Idaho has been built on the peculiar skill and knowledge of the Basque sheepherders and that he knows of no case in which men have been trained for this work in this country and no case in which native-born Americans have entered this industry as sheepherders."[49]

Thus, as part of a common procedure among western woolgrowers, the NRSOA aimed to expound before the federal government the scale of the labor problem at a regional level. Eventually, Karl King and John Landa were able to clarify the situation and the Basques were admitted on visitors' visas for a period of six months (thus, every six months they had to renew their permits). Because the imported three hundred Basques were on six month's temporary permits, woolgrowers with the assistance of Karl King were required to obtain a new certification every six months. Due to this uncertain situation, King stated that "it will be necessary to convince the United States Employment Service and the Immigration Service that their [Basques'] services are needed and that they cannot be replaced by American labor."[50]

The labor agreement with the Spanish regime required of ranchers the following: to register their workers at the Spanish consulate in San Francisco; to submit all the legal documentation; to deduct $50 per month from the wages of each man for a period of one year, or a total of $600 for a period of five years (at which time the individual may withdraw his $600). Furthermore, the contract established a minimum wage at $150 a month.[51]

By that time, there were pending in Congress five separate bills which had as their aim the establishment of permanent residence in the United States for these Basque immigrants and those who had entered previously. In addition, there was a bill pending which had passed the House and was being held up in the Judiciary Committee of the Senate which provided for the exclusion of all Basques at that time in the United States on temporary permits. King remarked that "the success of the new program will depend entirely upon the cooperation given by the sheep growers to the program and through their cooperation with the United States Employment Service." By May 1948, King had received a new request from several individual sheep ranchers of Nevada, California, Idaho, Utah, and Nebraska to bring in an additional group consisting of 125 Basques. Moreover, King received a request to bring in 75 Greek immigrants to work for Greek-owned sheep outfits in southwestern Colorado and southeastern Utah.[52]

In August 1948, the Spanish regime drafted a new labor code applying exclusively to the Basque immigrants that aimed to increase regulations on the recruitment of these workers. Thus, this regulation benefited Franco's regime by providing revenue. Apparently, the new labor agreement intended to improve Basques' working conditions and defend their interests in the United States. The agreement read as follows: "The Government of Spain is desirous of protecting the interests of its nationals, and in the interest of protecting and safeguarding the interests of the Spanish nationals, parties of the second part, and to provide a working understanding with parties of the first part." With this contract, the western sheepmen were forced to accept the conditions that they pay all the immigrants' expenses and to establish a minimum wage at $175 a month. In addition, they should be entitled to the benefits of any improvement in wages and working conditions, which may be instituted and prescribed during the period of employment. They should also receive the benefits and advantages of any social welfare in the United States. Employees must be employed exclusively as sheepherders in the states of Nevada, Idaho, Utah, Wyoming, Colorado, California, Arizona, Oregon, and Washington. All disputes between employees and employers must be resolved according to the federal labor laws and no deduction must be made from the wages of any of the employees hired under this agreement for commissions, fees, or any other purposes. The employer must permit sheepherders to visit nearby towns for the purpose of making necessary purchases, receiving medical care, or as a vacation, without deductions in vacation pay.[53]

Moreover, Franco's regime used the same discourse as ranchers in favor of the Basques in order to justify the regularization on the recruitment process. The agreement read:

> The Spaniards [Basques] have proven their ability to raise and produce lambs and save the twin lambs, and because of this the employee is to receive an additional compensation . . . as follows: 25 ¢ per head on all lambs raised and produced over and above 95% of the total number of ewes in the band of sheep being herded by the employee when lambed on the open range and 10 ¢ per head over and above 95% when lambed on the ranches and in sheds.[54]

The Spanish government used the demand for Basque immigrant labor for sheep grazing in the United States as another channel to expand commercial relations with the United States. Because the new employment-related regulation from the Spanish regime increased the regulatory burden and placed additional costs on employers, Karl King expressed his disagreement with the Spanish authorities about the new contracts for Basque sheepherders and renegotiated the labor agreement, making the worker himself liable for his debts and without compensating according to the growth rate of labor productivity, among other things. In mid-October 1948, the contract was forwarded to the Spanish central government in Madrid for consideration.[55]

The winter of 1948–1949 was one of the most severe on record in the Great Basin. In January 1949, continuous heavy snowfall, severe ice storms, and bitter winter weather caused considerable hardship for the livestock industry of Nevada. Cattle and sheep were starving to death when thousands of animals were blocked up with snow. By late January, in eastern Nevada, at least 100,000 sheep and 20,000 cattle were marooned. In Ely, Nevada, the Basque sheepman Beltran Paris remembered later: "That winter we had over $100,000 in extra expenses. We lost those 200 sheep and 150 cows, too." Livestockmen feared losses that ran into millions of dollars and asked their representatives to declare a state of emergency in Nevada. From Washington, D.C., Patrick McCarran personally coordinated and led the United States Air Force in "Operation Haylift," an emergency airlift dropping more than five hundred tons of hay to feed these flocks. McCarran further pledged to allocate funds for immediate emergency relief for Nevada ranchers.[56]

On January 24, 1949, while many Nevada sheep ranchers were suffering great losses during this terrible winter, it was reported in the Senate a bill "for the relief of certain Basque aliens." On February 8, the

Senate approved the bill. Nevada sheep operators and those from other western states expressed their joy. In its March 1949 issue of the bulletin, the Idaho Wool Growers Association declared: "We are very pleased to announce the United States Senate has approved a bill which legalizes the entry of certain Basque alien sheep workers that came to the United States in 1944."[57]

Welcome, Mr. McCarran!

In the immediate postwar period, Pat McCarran appointed himself mediator in the diplomatic relations between the United States and Spain. In his attempts to strengthen relations with Franco's Spain, McCarran aimed among other things to resolve the question of Basque immigrant labor in Nevada. McCarran's intensive lobbying to establish a stable source of Basque immigrant workers opened new windows in the international relations between the United States and Franco's Spain. The senator from Nevada was a key figure in the negotiations and parallel recruiting process.[58]

In July of 1949, Patrick McCarran had asked the Senate to grant an economic aid package of $50,000,000 to Spain. He lobbied to supply increasing financial assistance to counter the Soviet Union. On May 5, 1949, McCarran, in his position as chairman of the Senate Appropriations Committee, threatened the State Department with a cut in its budget unless it showed a friendlier attitude toward the dictatorship of Francisco Franco. The senator from Nevada publicly warned Secretary of State Dean Acheson that until the American policy toward Spain is changed, "I am going to look into this appropriation with a fine-tooth comb." In July, President Truman opposed aid to Franco's regime and the loan of $50,000,000.[59]

In the fall of 1949, Patrick McCarran used his senatorial power to frustrate every attempt to liberalize and amend the existing Displaced Persons Act. He found ways of evading his senatorial responsibilities in his efforts to obstruct the amending of the displaced persons law. On the pretext of carrying out a survey of the administration of the Displaced Persons Act and expenditures of European Recovery Program funds, McCarran asked for a three-week leave of absence to go Europe. The request was granted and within a few days, on September 14, Pat McCarran and his wife Birdie departed from the port of New York to Cherbourg, France, on the Cunard White Star liner *Queen Mary*. In New

York, before sailing to Europe, McCarran notified the press that he intended to meet with Francisco Franco to discuss with him the issue of diplomatic recognition for Spain and the possibility of a loan. McCarran said that he was going "to meet the Spaniard in his own back yard, so to speak" and he wished to "talk to every member of the Spanish Administration, including Mayors of principal cities, to find out what's going on." The next day, President Truman distanced his administration from the Nevada senator's visit to Spain.[60]

McCarran traveled around Europe interviewing different commissioners in his attempt to investigate administration of the Displaced Persons Act and expenditures of the European Recovery Program (the Marshall Plan). Besides France, McCarran and his wife traveled to Germany, Belgium, Switzerland, and Italy. Afterward, he had an audience with Pope Pius XII at the Vatican. On October 30, McCarran crossed the border into Spain. By Monday, October 31, he arrived in Madrid.[61]

Two days later, on November 2, in the morning, McCarran met Francisco Franco at the El Pardo Royal Palace, the dictator's official residence in Madrid. During this meeting, McCarran advised Franco to exploit its anticommunist stance in order to find possible ways to open relations with the United States. Further, he shared interests such as a strong Catholic faith, virulent anticommunism, and passion for hunting. In the meeting, also, McCarran probably mentioned the question of the Basque sheepherders. He was received gladly by the Spanish authorities and was seen as the best American friend of Franco. Next day, the Spanish monarchic newspaper of Madrid *ABC* covered the visit of the American senator, publishing a large photograph of the meeting between McCarran and Franco and celebrated how Franco "welcomed. . . . Mister Mac Carran [*sic*]." Four years later a Spanish movie would appear directed by Luis García Berlanga with the title *Welcome, Mr. Marshall!* The theme emphasized a much-anticipated visit by a group of American emissaries with their promises of economic aid. Circumstances were similar to the earlier visit of McCarran and the hopes that he represented for future improved relations and aid from the United States.[62]

The meeting went well. McCarran told American reporters that Spain had a democratic system of government and further said that Franco had "very democratic feelings." He was impressed by Franco's "great alertness for world affairs." Contrary to Spain's international image after 1939, as a place of a highly repressive and authoritarian regime, McCarran contended that Franco's dictatorship enjoyed broad support by the Spanish people. On the following day, taking advantage of his visit to Madrid, McCarran, accompanied by Thomas E. Dewey, the 1948 defeated Republican presidential candidate, met Martin Artajo, Spanish Minister of Foreign Affairs, at the Santa Cruz Palace to discuss in detail economic aid and possible strategic military negotiations with the United States. On November 4, from Zaragoza, Spain, McCarran told reporters that it was "absolutely necessary Spain be admitted into the Atlantic pact." Because of its geostrategic position, Spain "must be treated as she deserves and must occupy a preferential place in the nation's community."[63]

From there, McCarran went to Paris and from there to Ireland, where he found his family's roots. There McCarran curiously or perhaps predictably came out in favor of the independence of Northern Ireland from Great Britain. On December 2, McCarran and his wife sailed from England to New York. Within five days, after more than two months of travel around Europe, the McCarrans were back in the United States.[64]

The American press reacted negatively to McCarran's visit with and sympathy for Francisco Franco. In political circles, the event stirred controversy. Particularly, Democratic liberals attacked McCarran for being overly friendly with Francisco Franco. In November 1949, columnist Drew Pearson writing for his "Washington Merry-Go-Round" column in the *Washington Post* criticized McCarran as a cynical senator who not only abused his powers for undertaking unofficial diplomatic initiatives with Franco, but further gave advice to the Spanish dictator on foreign affairs. Pearson said: "Senator Pat McCarran of Nevada, No. 1 enemy of Europe's homeless refugees, has now set himself up as chief volunteer public relations counsel to Europe's No. 1 Fascist dictator."[65] In the same stratum, political commentator Robert Allen criticized McCarran as an "ardent Franco supporter" because he was traveling overseas not only as a watchdog for the administration of the displaced persons law but also as a private emissary to the Spanish dictator.[66]

Some Nevada newspapers took editorial note of McCarran's visit to Franco.[67] The *Reno Evening Gazette* said that despite McCarran's good impressions of the Spanish dictator, "he still represents the Fascist type of dictator, and his friendship for Hitler and Mussolini is not forgotten. The persecution of political enemies and discrimination toward minority religious groups have not abated." The November 11 *Reno Evening Gazette* editorial printed a picture of Pat McCarran and Francisco Franco at their meeting at El Pardo in Madrid. McCarran was not the only American representative visiting Franco in the late 1940s. For example, in September 1949, James J. Murphy, a Catholic representative from New York, and U.S. Navy Admiral Richard L. Conolly visited the Spanish dictator. McCarran's visit to Franco must be understood in the context of an increasing number of contacts with the United States in 1949.[68]

McCarran's trip to Europe coincided with an intensification of the Cold War. In August 1949, the Soviet Union detonated its first atomic bomb and on October 1, the Communist People's Republic of China proclaimed victory over the American-backed nationalist forces of Chiang Kai-shek. Both events seemed to confirm McCarran's argument about the need to grant economic assistance to Franco's Spain to enlist him in the worldwide struggle against communism. In the late 1940s, with the worsening of the Cold War, a growing number of representatives began to adopt a hardline nationalist stance to any major international event. Numerous voices appeared more flexible regarding the undemocratic nature of the Francoist regime and expressed a willingness to accommodate the Spanish fascist state in the Cold War logic.[69]

McCarran's visit to Franco lifted his hopes for American economic aid and diplomatic recognition. At the same time, the Spanish regime showed increasing interest in the Basque sheepherder issue. Franco listened to McCarran's advice and embraced a more anticommunist western-oriented position. In retrospective, McCarran's trip represented the prelude of an American diplomatic approach to Spain. For McCarran, it served to establish good relations, which in turn helped facilitate the procedures to contract Basque immigrants for the sheep industry. In the late 1940s, the American press regarded McCarran as a staunch friend of the Spanish dictator.

On the other hand, Franco's regime had no clear plan for handling the issue of the Basque sheepherders. Previously, it had accommodated American commercial and business interests. The continued demand for Basque sheepherder immigrants in the American West forced the regime to pay more attention to the recruitment process. As it dealt with this question, the regime sometimes sought leverage in its efforts to cultivate American friendship. Others adamantly opposed Basque emigration, considering it a loss for the domestic economic recovery and demography.

In the Spanish administration, there were some influential men who continued to raise voices against any commercial concessions to the United States. In response to the labor agreement between Karl King and the Spanish authorities, on February 11, 1949, Carlos Manzanares, minister of foreign affairs, sent to the Ministry of Labor an unfavorable report about the Basque immigration in the American West. Manzanares described the recruitment of Basques as a labor "traffic," from which the sole beneficiaries were certain livestock private companies and operations. He said that the U.S. government was not interested in encouraging "this kind of immigration." Manzanares argued that the Spanish regime did not benefit much from this emigration, because most of the Basques did not return and those who did, did not bring significant funds back to the Spanish economy. He further wrote that the emigration in terms of numbers was negligible, but he did observe that a loss to Spain because the "Basque-Navarre man tends to be physically and spiritually healthy." In addition, he said that immigrants were culturally uprooted when resettling in the American West. Manzanares remarked that Basque sheepherders were highly vulnerable to economic exploitation in those western rural remote places and with their ignorance of the English language that "put them at the mercy" of their employers. He concluded that various administrative regulations designed to protect them were insufficient in this process of "Recruitment, Traffic or Importation of Sheepherders."[70]

Carlos Manzanares suggested first stopping Basque emigration to work in the sheep industry of the West under such conditions. Moreover, he asserted the need to defend the interests of those Basques living already in the United States. Manzanares prohibited Karl King from undertaking additional recruitment practices in Spain. He ordered the Spanish consul in San Francisco to carry out a detailed investigation of the latest applications through the *Carta de Llamada,* intending to determine possible fraud cases. Manzanares then asked to inform the Spanish Embassy in Washington, D.C., and inform every consulate in the United States.[71]

Dealing with Franco's Spain: Opening Dialogue, Recognition, and Consolidation of Agreements, 1950–1955

On February 9, 1950, senator Joseph McCarthy from Wisconsin gave his famous speech in Wheeling, West Virginia, claiming that over two hundred communists were infiltrated into the State Department. McCarthy's speech alarmed the whole country and influenced national politics. As the anticommunist movement gained momentum in the context of the Cold War, this cause increasingly became institutionalized and professionalized with the consent of the federal government.

For the next four years, Joe McCarthy worked persistently to expose "subversives" in the United States. In 1950, the Tydings committee, formed to investigate McCarthy's charges, concluded that his allegations were "a hoax and a fraud." McCarthy's allegations, however, influenced American public opinion within a troubled international context. The rise of McCarthyism was paralleled by a drastic deterioration of the Cold War abroad and at home. In June, the United States intervened in the Korean War and shortly after, Julius and Ethel Rosenberg were arrested and charged with espionage activities against the United States.[72]

McCarran soon joined McCarthy in the anticommunist crusade, sometimes getting more visibility than McCarthy himself. In late summer 1950, as the American troops deployed on the Korean Peninsula, the Senate approved a loan of one billion dollars to Franco's Spain under the administration of the European Recovery Program. This indicated the United States' interest in pursuing Spain as an important geostrategic territory in Western Europe. While President Truman and a group of Democratic representatives denounced this loan, Franco and his officials welcomed the development. Subsequently, in September 1950, diplomatic negotiations accelerated between Spain and United States. Ángel Viñas has said that September 1950 marked a historical milestone in Franco's long effort toward the recognition of his regime by the United States.[73]

Earlier that year, on August 10, 1950, McCarran had introduced his anticommunist bill S. 4037 ("McCarran Omnibus Bill") or Internal Security Act, which aimed at tightening control over the communist presence in the United States. It provided provisions requiring registration of all communist-front organizations and individuals, which had been vaguely defined. This bill endowed the FBI with greater powers to conduct investigations and the capacity for repressive actions against communists. In September 1950, Congress passed the McCarran Internal Security Act, over President Truman's veto. At the same time, on August 23, the Senate passed another bill granting permanent residence to 163 Basque sheepherders living in the United States. Within the last years, Pat McCarran's health had increasingly deteriorated, after suffering several heart attacks in 1946 and 1947. Despite his failing health,

McCarran kept his position in the Senate and actively worked to take action against the alleged communist threat. He supported every piece of legislation that in one way or another was aimed at discovering "subversive activities."[74]

Again in 1950, McCarran (along with Senator Francis Walter from Pennsylvania) introduced a bill in the Senate that provided a set of mechanisms to exclude communists from immigration and naturalization in the United States. In 1952, after a long debate, the Congress passed the Immigration and Nationality Act, known as the McCarran-Walter Act.[75] It focused on the deportation of "undesirable aliens." This bill, while it did not eliminate the quota system, established the principle of preference categories, giving priority to skilled immigrants and to family reunification. At that time, Oscar Handlin, as a contemporary historian, wrote: "Out of his own fears and the fears of others, capitalizing on his key part in a touchy political situation, McCarran assembled the forces that brought him easy victory in the struggle to determine the future immigration policy of the United States."[76]

"A Big Milestone in the Work of the Sheep Raisers of the West": The Sheepherder Bill of 1952

In Fresno, California, a group of Basque Americans involved in the sheep business founded in 1950 the California Range Association (CRA), arising from the California Wool Growers Association, to work for the importation of needed Basque sheepherders into California and the West on a permanent quota basis. The CRA assumed the role hitherto fulfilled by Dangberg's NRSOA, by coordinating the supply and demand for Basque immigrant labor among all the western woolgrower associations. The first members of the board of directors were: John P. Bidegaray, president; Philip Erro, vice president; John Irola, Jose Azparren, and Pete Choperena.[77]

On August 10, 1951, the board of directors of the CRA appointed a special committee to consider a plan along with the western congressional delegations to address the Basque labor question. The special committee working on the proposed movement included: E. B. Elorriaga, Fermin Huarte, and Ramiro Azparren. One of their objectives was to seek legislation to bring in increasing numbers of Basques. In late 1951, both NRSOA and the CRA had joined together in this process. On October 8, Robert Franklin, secretary of the CRA, traveled to Washington, D.C., to discuss with congressional representatives and government officials a possible plan to bring more Basques from Europe.[78]

On January 30, 1952, responding to the western ranchers' demands, Patrick McCarran introduced the bill S. 2549 in the Senate in order to "provide relief for the sheep raising industry by making special quota immigration visas available to certain alien sheepherders." The bill guaranteed permanent employment for prospective "skilled" sheep workers by issuing a special immigration visa. On February 25, the Senate passed the sheepherder bill and on April 2, it was approved by the Congress.[79]

While the bill awaited signature on President Truman's desk, the *National Wool Grower* magazine in its April 1952 issue greeted the news with unrestrained joy. It remarked that both the cooperation among all western woolgrower associations and the work of the congressional delegations were crucial in obtaining this result:

> The passage of S.2549 is not only an example of the value of the close cooperation between the State and National wool grower groups on such matters but also demonstrates the faith of the Congress in the wool growers' organization. The Senate Committee's report approving the measure pointed out that the intent of the Congress was that sheepherders coming in under the bill must have the approval of established wool grower groups.[80]

On April 12, President Harry Truman signed the sheepherder bill. As it was expected, the CRA announced happily: "PRESIDENT SIGNS SHEEPHERDER IMPORTATION BILL! [*sic*]" In an urgent written notice, John P. Bidegaray announced to their members that the president had finally signed bill S. 2549, which permitted the importation of five hundred skilled sheepherders into the United States. The "Sheepherder Bill" made special quota immigration visas available to certain alien sheepherders, or better said, to Basque sheepherders. Essentially, this act was directed at stabilizing and facilitating the recruitment of a Basque labor force to be employed in the sheep industry of the American West. Bidegaray said that this was "a big milestone in the work of the sheep raisers of the west to obtain needed manpower." In the summer of 1952, the CRA proceeded to arrange everything required for securing the Basque immigrant workers.[81]

The western woolgrower associations also saw advantages in the Immigration and Naturalization law also introduced by McCarran. They knew that if the McCarran-Walter Act was passed, in addition to the special legislation to admit the five hundred Basques, it increased the immigration quota by 50 percent from each country—Spain and France. The National Wool Growers Association, California Wool Growers Association, California Range Association, and other state woolgrower organizations supported this bill vigorously. Since Basques were considered "skilled" workers (sheepherders), they received special treatment or admission status under the provisions of the new Immigration and Naturalization Act.[82]

To many, McCarran's enthusiasm for importing Basque immigrants stood in contrast to his general immigration policy and particularly clashed with those supporters of the Jewish refugee resettlement issue in the aftermath of the war. Basques became a major subject in the political debate concerning the McCarran-Walter Act. From Washington, D.C., Peter Edson, correspondent for the Newspaper Enterprise Association, noted the inconsistency in McCarran's position on the issue of Basque sheepherders. In March 1951, he wrote an article entitled "McCarran Act Backfires on Its Sheepish Author." He criticized McCarran's efforts to recruit Basque immigrants as merely serving the narrow interest of his constituents: "The stiff, controversial McCarran act intended to keep subversive immigrants out of the United States and control those who are already in, has now backfired against the author of the act himself, Sen. Pat McCarran of Nevada. It's all because of those Basque sheepherders Senator McCarran wanted to import for the benefit of his Nevada sheep-raising constituents."[83]

Edson complained about McCarran's ethnic-preference policy that privileged Basque immigrants (subjects of a totalitarian government) contrary to the provisions of the McCarran Immigration Act that discouraged immigration from such governments.

Peter Edson vaguely defined Basque immigrants as followers of Francisco Franco and connected them with the Spanish dictatorship by saying that they were Spanish *Falangists*.[84] This journalist added that Basques "might have been admitted on temporary visas if they could prove no official connection with the Spanish Falange." Contrary to the previous judgments that saw Basques as possible communists, this American journalist considered Basques as subjects of Francisco Franco's dictatorial regime in Spain. Under the McCarran Act, as he correctly pointed out, they should probably be excluded. In response to these accusations, Henry Dworshak, senator from Idaho, one of the western representatives defending recruitment of Basque immigrants on permanent basis, declared that Basques fought against both the *Falangists* and the communists.[85]

In the early 1950s, despite the liberal criticism, the legislation designed to import small groups of Basque immigrants and the establishment of a political lobby group, designed to foster the sheep industry, also helped to promote a Basque identity in the American West that embraced ideas of hard work, perseverance, and special talents for sheep husbandry. The formulation and passage of the act served to reinforce the idea that Basques represented an exceptional and anomalous racial stock in the "scale" of being "white," making them a more desirable immigrant. Many observers reinforced the idea that the Basque was a distinguishable "race" with natural skills for sheep grazing practices.

On May 12, 1952, *Life* magazine published a six-page photographic essay on the Basque sheepherders in the Great Basin. The article showed the work on a Utah sheep ranch during the lambing season through the experience of Bertrand Borda, a twenty-year-old Basque newcomer in the Interior West. This essay idealized Basques' work by presenting them as an uprooted immigrant group with labor skills for sheep grazing who were transplanted into an unfamiliar environment: "With a few helpful tips from his fellow herders and with his native steadiness and inbred ability to lead the life of his flock, Bertrand learned his job quickly." Further, it read that Basque sheepherders were highly devoted workers and "newborn lambs are almost sacred" to them. This essay was remarkable publicity for the Basques in the United States.[86]

On November 23, 1952, the *New York Times* published in its Sunday magazine edition an article entitled: "Basques Are Best." While describing the 1952 sheepherder bill, the author said Basques were recognized as the unique people with whom western sheep operators could entrust their flocks: "United States sheep owners discovered several generations ago that the Basques possessed the skill and moral qualities for that job." It described Basques as honest, self-reliant, sturdy, shrewd, and honorable people. The article assumed that the alleged Basques' skills working as sheepherders were because they were already socioculturally preadapted for the extensive sheep grazing conditions and even to the new environment in the North American West. To the author, the natural environment in which Basques lived in the Old World explained their skills in managing sheep: "Virtually isolated from the outside world for thousands of years, the Basques tilled the soil and raised sheep and cattle." The article stressed that their racial traits and romantic culture made Basques master sheepherders. It contended that Basques are the world's finest sheepherders because of their intensive labor and devotion to the ovine animals:

> A Basque will not only stay with the flock through intense heat or a blizzard, and die with it rather than save his life, but he will also give his charges the most conscientious and skillful care. While other shepherds may just take the flock out to graze and limit their job to watching passively day after day that no sheep is lost, the Basque makes sure that the ewes do not become overtired, that they have the proper feed for the morrow close at hand and that the grazing ground is free of cockleburs and other grass that clings to the wool and lowers its quality because it must be combed out after the shearing, which means a waste of time and money. As a result, ewes entrusted to Basques are generally fatter, stronger, give richer milk and produce better lambs.[87]

Besides having intrinsic talents for sheep grazing, the author said Basques were honest sheep workers, an important attribute for immigrant workers. Most important was that they were "dependable herders."[88]

Closing a Chapter, 1953–1955

By 1953, Pat McCarran received applause from his constituency for his work for Nevada and was honored by different groups across the country. During the weekend of September 19–20, in conjunction with the Washoe County Fair, the state of Nevada celebrated "McCarran Day" in Reno. At noon on September 19, as a prelude to the main festivities, over 250 personalities assembled at a luncheon celebration held in the Sky Room of the Mapes Hotel to honor the senator. Celebrating McCarran's golden wedding anniversary, his fifty years of public service, and his seventy-seventh birthday, different organizations and individuals of Nevada offered praise and presented gifts to McCarran. Robert Franklin, representing the western woolgrower community, was among the well-wishers who praised McCarran's efforts in the recruitment of Basque immigrants. Next to Franklin was Ruthio Iturri, a Basque sheepherder in Nevada, who said some words in Basque language in praise of McCarran for working on behalf of the Basque-American community in Nevada and the West. Iturri presented him with a *bota*, the traditional leather wineskin container.[89]

In late summer 1953, under threatening international circumstances, the U.S. government saw the crucial importance of Spain in the geostrategic containment of the Soviet Union. On August 25, 1953, Francisco Franco decorated Senator Patrick McCarran with the *Gran Cruz de Isabel la Católica* (Grand Cross of Isabel the Catholic) for helping to improve relations between Spain and the United States. In late September, the United States recognized Franco's dictatorship by signing the Pact of Madrid, the first military-economic agreement between the two governments. On September 26, U.S. ambassador James C. Dunn and Martin Artajo signed three bilateral agreements: the first agreement dealt with the construction of U.S. military facilities in Spain; the second referred to the financial aid to Spain; and the third agreement set forth the terms of the military assistance to Spain.[90]

The Pact of Madrid ended a period of international isolation of the Franco regime. The bilateral relationship reinforced domestically and internationally the Francoist dictatorship. The United States, for its part, strategically installed military bases in Spain, gaining maneuvering space in its foreign policy against the Soviet Union. In the context of the negotiated agreements, Franco's regime appeared increasingly willing to adopt laxer labor laws and make easier the recruitment process at great cost to the Basque immigrants' labor conditions. Nonetheless, most of the Basques were willing to accept any labor conditions in the United States in exchange for a passport out of Spain. This was the case of Peter Arraras, who worked for some years in Nevada as a sheepherder and returned to Spain. In 1952, however, he appeared impatient to go back to the Great Basin, wanting to escape from Franco's dictatorship. Arraras became one of those five hundred Basques allowed to travel and work in the United States under Bill S. 2549 that Truman signed in 1952. Envious of the Basques, other people from different provinces in Spain, such as Huelva, addressed the Spanish immigration officials desiring to immigrate to the United States to work as sheepherders as well. At this point, the central government in Madrid recognized the importance of the Basque question for the improvement of relations with the United States. The Ministry of Labor now enlarged the numbers Spain would allow to emigrate to the United States.[91]

On February 1, 1954, in the Eighty-Third Congress, McCarran introduced another bill to "provide relief for the sheep industry by making special nonquota immigration visas available to certain skilled alien sheepherders." The bill offered both permanent employment for "such skilled alien sheepherder" and admission for legal permanent residence in the United States without being subjected to the quota treatment applied to Spain.[92]

Although it did not make any reference to the Basques, McCarran obviously introduced this bill in the Senate having in mind Basque immigrants. With "skilled alien," McCarran meant Basques. The senator from Nevada intended to bring an additional group consisting of 385 Basques. However, by the time the western woolgrowers received the legal authority to import Basques, the sheep industry in the West was facing challenges beyond the labor question. The future of the sheep industry faced variables in tariff regulation, weakening market demand, and public-land regulations. On September 3, 1954, President Dwight D. Eisenhower signed McCarran's latest bill on behalf of the Basques.[93]

Less than a month later on September 28, 1954, McCarran collapsed and died of heart attack while addressing a Democratic rally in Hawthorne, Nevada. Unsurprisingly, on September 30, the Francoist newspaper *ABC* published a one-page article praising McCarran for his efforts on behalf of Franco's dictatorship and lamenting his loss: "In his death, Spain has lost a grand and sincere friend" [writer's translation]. It further praised his public career, saying: "Pat McCarran was always characterized by his tenacity . . . [and] extraordinary spiritual brightness. [He was a] convinced and impassioned anticommunist." On December 2, the Senate censured Joseph McCarthy for his anticommunist demagoguery.

Despite the hard fought importation of additional Basque workers, by the time the Sheepherder Bill was passed, the open-range sheep industry in Nevada and the West was struggling to survive in the postwar economy. Also, in November 1954, Ellis Island, the major port of entry for Basque immigrants that had opened in 1892, closed its doors. Now, Basque immigrants, like other groups, arrived in American airports. This was another indication that vast technological changes were taking place and so too was the American economy and society. Many of these changes would eventually weaken and undermine the sheep industry in the American West.[94]

Summary

While the United States and Spain maintained chilly political relations in the early postwar period, they forged economic and military links in the face of the realities of the Cold War. As the Soviet Union became a threat to the United States in the Cold War, Franco's Spain used and reinforced its anticommunist ideology to advance its foreign relations with the United States and obtain economic aid. Amid the changing international situation, the question of Basque immigration to the United States drew the attention of Franco's regime.

By the late 1940s, the Spanish government manipulated the Basque immigration issue for its own political and economic advantage. Spain's government began regulating Basque immigrant labor to the United States through the enforcement of immigrant contract-labor laws. The Spanish legislation required American ranch employers to raise wages and improve the conditions of sheepherders' work. Franco's government imposed taxes on Basque immigrant workers that were hired in Spain to work in the American West. The increasing administrative red tape raised the cost of recruiting Basques from Spain. The 1953 Pact of Madrid between Spain and the United States also promoted other commercial relations and indirectly facilitated the recruitment of Basque immigrant labor.

By the early 1950s, as the passage of the 1952 sheepherder bill reflected, Basque immigrant labor became institutionalized. Basques were officially given a privileged position and almost publicly endorsed as natural-born sheepherders. In the early 1950s, Basque immigrants entering sheep grazing escaped from Franco's Spain, benefiting from the reputation that had already been created and preceded them in sheep grazing. The sheep industry at the same time was adjusting to changing economics after the war. By the mid-1950s, various new developments threatened the sheep industry in Nevada and the American West. Changing national market outlooks, increasing international competition, popularization of new synthetic fibers, and stricter regulation of public lands pushed the sheep industry into chronic economic crises.

Notes

1. Declaration of Intention, Rafael Fagoaga, no. 901, April 30, 1952, State District Court, Elko County Courthouse, Elko, Nevada; Petition for Naturalization, Rafael Fagoaga, no. 677, May 19, 1958, State District Court, Elko County Courthouse, Elko, Nevada; Obituary, "Jean Pilar Fagoaga," *Elko Daily Free Press*, December 8, 2015.
2. "Basques Have Good Qualities," *Walla Walla Bulletin*, March 4, 1945; by then, also, Edward Norris Wentworth, the distinguished animal husbandman graduate of Iowa State College and historian of the sheep industry, identified Basques as one of the major "racial" groups traditionally involved in the sheep industry

of the American West, who spoke their language and "retained the *mores*" of their native country. Edward Norris Wentworth, "Eastward Sheep Drives from California and Oregon," *Mississippi Valley Historical Review* 28, no. 4 (March 1942): 520–21; Charles W. Towne and Edward N. Wentworth, *Shepherd's Empire* (Norman: University of Oklahoma Press, 1945), 268; by 1948, Wentworth completed his voluminous *America's Sheep Trails*, a chronicle that comprises the history of sheep husbandry in America.

3. Elliott, *History of Nevada*, 312–18; Foss, *Politics and Grass*, 84–98; *Idaho Wool Growers Bulletin* 24, no. 21 (October 17, 1945); *Idaho Wool Growers Bulletin* 25, no. 13 (June 26, 1946); *Idaho Wool Growers Bulletin* 25, no. 21 (October 16, 1946); *Idaho Wool Growers Bulletin* 25, no. 24 (November 27, 1946); *Idaho Wool Growers Bulletin* 25, no. 26 (December 25, 1946).
4. Affidavit, John Dangberg and D.W. Park, September 29, 1945, State District Court, Douglas County Courthouse, Minden, Nevada, Box BSQAP 0114, Nevada Range Sheep Owners Association Papers, 1944–48, Jon Bilbao Basque Library.
5. Ibid.
6. Ibid.
7. John Dangberg to Harmon E. Hosier, October 1, 1945; Harmon E. Hosier to the NRSOA, October 3, 1945; John Dangberg to the Commissioner of Immigration and Naturalization in Philadelphia, October 4, 1945; John Dangberg to the Commissioner of Immigration and Naturalization in Philadelphia, November 5, 1945; Harmon E. Hosier to the NRSOA, December 12, 1945; Arthur N. Corcoran to the NRSOA, March 6, 1946; John Dangberg to Ángel Sanz Briz, March 21, 1946, Box BSQAP 0114, Nevada Range Sheep Owners Association Papers, 1944–48, Jon Bilbao Basque Library.
8. These immigrants included: Cirilo Olavarri, Theodoro Asla, Antonio Urteaga, and Joe Buera all of them settled in Utah; and Vidal Mezo, Marcial Aguirregoitia, Geronimo Bilbao, Juan Tomas Mendiola, Juan Pedro Eguibegui (alias Raymond Etchevers), Miguel Iriarte, and Alejo Yraguen, all of eastern Nevada. H.R. 1402, Private Law 818, 79th Cong., 2d Sess., August 7, 1946, *Calendars of the United States House of Representatives and History of Legislation, Seventy-Ninth Congress, Final Edition* (Washington, D.C.: Government Printing Office, 1946), 69, 113; this law was preceded by others: S. 314. H.R. 1944, Private Law 286, 77th Cong., 2d Sess., February 19, 1942, *Calendars of the United States House of Representatives and History of Legislation, Seventy-Seventh Congress, Final Edition* (Washington, D.C.: Government Printing Office, 1942), 93, 113, 202, 247; H.R. 2626. 78th Cong., 2d Sess., December 5, 1944, *Calendars of the United States House of Representatives and History of Legislation, Seventy-Eighth Congress, Final Edition* (Washington, D.C.: Government Printing Office, 1944), 108, 193.
9. Felipe Albizua Calzade, Pedro Olano Filibi, Felix Lezamix Gueregueta, Julian Muruaga Ispizua, Golo Zuazua Izaguirre, Fermin de Bilbao Jayo, and Teodoro Basabe Uriarte. John Dangberg to Ivan Williams, October 7, 1946; John Dangberg to Ivan Williams, October 15, 1946; John Dangberg to Ivan Williams, October 17, 1946; John Dangberg to Ivan Williams, October 24, 1946; D.W. Brewster to

Arthur H. Bohoskey, November 18, 1946, Box BSQAP 0114, Nevada Range Sheep Owners Association Papers, 1944–48, Jon Bilbao Basque Library.

10. A.E. Lawson to John Dangberg, November 20, 1946; John Dangberg to J.M. Jones, November 27, 1946; John Dangberg to M.C. Claar, November 30, 1946; John Dangberg to G. Mendiburu, November 30, 1946; John Dangberg to A.E. Lawson, November 30, 1946; J.M. Jones to John Dangberg, December 3, 1946; M.C. Claar to John B. Dangberg, December 7, 1946; John Dangberg to M.C. Claar, December 13, 1946; M.C. Claar to John Dangberg, December 16, 1946, Box BSQAP 0114, Nevada Range Sheep Owners Association Papers, 1944–48, Jon Bilbao Basque Library.
11. Joseph Savoretti to J.M. Jones, December 16, 1946, Box BSQAP 0114, Nevada Range Sheep Owners Association Papers, 1944–48, Jon Bilbao Basque Library.
12. Edwin E. Marsh to State Association Secretaries, and Executive Committee, December 23, 1946, Box BSQAP 0114, Nevada Range Sheep Owners Association Papers, 1944–48, Jon Bilbao Basque Library; "Address, T.H. Gooding, President, to the Fifty-Fourth Annual Convention of the Idaho Wool Growers Association," *Idaho Wool Growers Bulletin* 26, no. 2 (January 22, 1947): 3.
13. John Dangberg to Patrick McCarran, January 20, 1947, Box BSQAP 0114, Nevada Range Sheep Owners Association Papers, 1944–48, Jon Bilbao Basque Library.
14. Edwards, *Pat McCarran*, 133; Ybarra, *Washington Gone Crazy*, 381–98.
15. Joseph Savoretti to John Dangberg, March 14, 1947; John Dangberg to Ángel Sanz Briz, March 19, 1947; Carlos Manzanares to John Dangberg, April 9, 1947; John Dangberg to Cipriano Aznaren, April 12, 1947; John Dangberg to Jose Saturnino Algarra de Carlos, April 12, 1947, Box BSQAP 0114, Nevada Range Sheep Owners Association Papers, 1944–48, Jon Bilbao Basque Library.
16. Members of the Nevada State Farm Labor Advisory Committee included: William B. Wright, Chairman; C.W. Creel, Director; Edwin Marshall, from Las Vegas; Norman Annett, from Wellington; Ed Settelmeyer, from Reno, "Resolutions of the Nevada State Farm Labor Advisory Committee," April 15, 1947; John Dangberg to L.J. Holland, April 16, 1947, Box BSQAP 0114, Nevada Range Sheep Owners Association Papers, 1944–48, Jon Bilbao Basque Library.
17. The questions asked were: 1. What are the names of the men now working for you that were imported by this association? 2. Is each of these men herding sheep? 3. Is the work of each satisfactory? 4. Does each man want to stay permanently in this country after March 14, 1948? 5. Will you need these men after March 14, 1948? 6. What will happen to your outfit if these men are taken away from you and returned to Spain after March 14, 1948? John Dangberg to the members of the NRSOA, October 6, 1947, Box BSQAP 0114, Nevada Range Sheep Owners Association Papers, 1944–48, Jon Bilbao Basque Library.
18. John Dangberg to M.C. Claar, October 6, 1947; John Dangberg to the members of the NRSOA, October 6, 1947; Reginald Meaker to Grace M. Dangberg, October 8, 1947; John Dangberg to Bert Ithurburu, October 24, 1947, Box BSQAP 0114, Nevada Range Sheep Owners Association Papers, 1944–48, Jon Bilbao Basque Library.
19. Tony Herrera to John Dangberg, November 16, 1947; Gordon Griswold to John Dangberg, November 9, 1947; John Dangberg to Mr. Osborn, October 6, 1947;

Mono Land & Live Stock Co. to John Dangberg, October 7, 1947; John Dangberg to the Uhalde Bros., October 7, 1947; Lyle L. Ellison to John Dangberg, October 11, 1947; Pete Borda and Raymond Borda to John Dangberg, October 13, 1947; Reginald Meaker to John Dangberg, October 20, 1947; J.W. Newman to John Dangberg, October 20, 1947; Bert Ithurburu to John Dangberg, October 23, 1947; John E. Robbins to John Dangberg, October 25, 1947; Alfonso Sario to John Dangberg, October 26, 1947; Andrew Little Jr. to John Dangberg, October 28, 1947; Emily Carricaburu to John Dangberg, October 28, 1947; Leonard Bidart to John Dangberg, November 6, 1947; L.J. Holland to John Dangberg, November 7, 1947; Al Bantley to John Dangberg, November 7, 1947; Ellis Ragan to John Dangberg, November 9, 1947; James Farmer to John Dangberg, November 10, 1947; W.W. Whitaker to John Dangberg, November 11, 1947; John Belaustegui to John Dangberg, November 16, 1947; C.E. Nicholson to John Dangberg, November 17, 1947; R.C. Rich to John Dangberg, November 20, 1947; Jean Uhart to John Dangberg, November 26, 1947; Robert Lloyd Pruett to John Dangberg, December 9, 1947, Box BSQAP 0114, Nevada Range Sheep Owners Association Papers, 1944–48, Jon Bilbao Basque Library.

20. These employers included: Andrew Little Jr. from Emmett, Idaho, contracted two herders; from Boise, James Farmer owner of the Desert Sheep Company contracted two; J.W. Newman from Twin Falls contracted two men; from Hammett, F.E. Wilson and Wilbur F. Wilson, heads of the family business Hammett Livestock Company, contracted two; and from the state of Washington, Ellis Ragan, owner of the Yakima Sheep Company, contracted two other Basques. M.C. Claar to John Dangberg, October 14, 1947; J.W. Newman to John Dangberg, October 20, 1947; Wilbur F. Wilson to John Dangberg, October 25, 1947; Andrew Little Jr. to John Dangberg, October 28, 1947; Ellis Ragan to John Dangberg, November 9, 1947; James Farmer to John Dangberg, November 10, 1947; M.C. Claar to John Dangberg, November 10, 1947; C.E. Nicholson to John Dangberg, November 17, 1947; R.C. Rich to John Dangberg, November 20, 1947, Box BSQAP 0114, Nevada Range Sheep Owners Association Papers, 1944–48, Jon Bilbao Basque Library.
21. Affidavit, John Dangberg and D.W. Park, December 10, 1945, State District Court, Douglas County Courthouse, Minden, Nevada; John Dangberg and D.W. Park to the Commissioner of Immigration and Naturalization Service in Philadelphia, December 10, 1947; J.H. Breckenridge to John Dangberg, October 11, 1947; Grace M. Dangberg to Eva B. Adams, December 10, 1947; Eva B. Adams to Grace M. Dangberg, December 16, 1947, Box BSQAP 0114, Nevada Range Sheep Owners Association Papers, 1944–48, Jon Bilbao Basque Library.
22. Martínez-Lillo, "La política exterior de España" 323–38; see also Florentino Portero, *Franco aislado: La cuestión española, 1945–1950* (Madrid: Aguilar, 1989); Carlos Collado Seidel, "¿De Hendaya a San Francisco? Londres y Washington contra Franco y la Falange (1942–1945)," *Espacio, Tiempo y Forma, Serie V, Historia Contemporánea* no. 7 (1994): 51–84; Paola Olla Brundu, "Europa y Estados Unidos frente al problema de la integración de España en el sistema de seguridad occidental (1945–1953)," *Spagna contemporanea* no. 15 (1999): 93–112.

23. David Caute, *The Great Fear: The Anti-Communist Purge under Truman and Eisenhower* (New York: Simon & Schuster, 1978), 177–78; Richard Gid Powers, *Not Without Honor: The History of American Anticommunism* (New York: Free Press, 1995), 134; Ellen Schrecker, *Many Are the Crimes: McCarthyism in America* (Boston, Mass.: Little, Brown, and Company, 1998), 73.
24. Ybarra, *Washington Gone Crazy*, 473.
25. Ronald E. Powaski, *The Cold War: The United States and the Soviet Union, 1917–1991* (New York: Oxford University Press, 1998), 72; Ernest S. Griffith, "The Place of Congress in Foreign Relations," *Annals of the American Academy of Political and Social Science* 289 (September 1953): 11–14.
26. To learn more about the diplomatic career of José Félix de Lequerica y Erquiza, see: María Jesús Cava Mesa, *Los diplomáticos de Franco: J. F de Lequerica. temple y tenacidad (1890–1963)* (Bilbao: Universidad de Deusto, 1989).
27. Theodore J. Lowi, "Bases in Spain," in *American Civil-Military Decisions: A Book of Case Studies*, ed. Harold Stein (Birmingham: University of Alabama Press, 1963), 675–76; Ángel Viñas, *En las garras del águila: Los pactos con Estados Unidos, de Francisco Franco a Felipe González (1945–1995)* (Barcelona: Crítica, 2003), 55, 57–62.
28. Ngai, *Impossible Subjects*, 236–37; John Dangberg to Patrick McCarran, January 10, 1948, Box BSQAP 0114, Nevada Range Sheep Owners Association Papers, 1944–48, Jon Bilbao Basque Library.
29. The bill contended that in the administration of the immigration and naturalization laws, the attorney general was authorized and directed to record the lawful admission for permanent residence of the following Basques: Jose S. Algarra de Carlos, Domingo Asiain Hualde, Patricio Asiain Hualde, Cipriano Aznarren Udi, Gracian Azparren Gazcue, Pedro Aspiroz Echechuri, Anastasio Barbarena Elizegaray, Melchor Burusco Domench, Pedro C. Cemborain Garmendia, Angel Dufur Iturri, Bernardo Dufur Iturri, Primitivo Egozcue Seminario, Modesto Elgorriaga Exposito, Felipe Errea Huarte, Esteban Erro Inda, Ambrosio Esnoz Recalde, Lino Goicoa Arozerena, Alberto Ibarregui Iriberri, Pablo Inda Sagardoy, Felipe Itturri Castilla, Francisco Jorajuria Andiaarena, Joaquin Juanarena Dufur, Francisco Larrea Ibarrola, Manuel Lecumberri Barber, Silvestro Martinex Exposito, Serpio Mendiguia Lerumbe, Jose Moulian Mendia, Jose Narvarez Berendiain, Antonio B. Nuin Exposito, Jose Ochandorena Petricorena, Jose M. Ochandorena Petricorena, Primitivo Olondriz Echeverria, Mateo Pedroarena Barberena, Leandro Urrutia Villanueva, Pedro Valencia Llorente, Prisco Villanueva Montoya, Marcelino Villanueva Urrutia, Salvador Vizcar Oroz, Demetrio Zubiri Uriz, Pedro M. Recaide Iribarren, Celestino Arozarena Elizagaray, Jose Betelu Zubiriaga, Dionisio Betelu Zubiriaga, Fermin Jorajuria Andiaarena, Tomas Iriarte Lerindegui, Miguel Redin Equiza, Miguel Inda Juandeaburre, and Jose Zubiri Coni, S. 1973, 80th Cong., 2d Sess., January 12, 1948; see also *Calendars of the United States House of Representatives and History of Legislation, Eightieth Congress, Final Edition* (Washington, D.C.: Government Printing Office, 1948), 174, 225.
30. John Dangberg to M. C. Claar, January 6, 1948; John Dangberg to Patrick McCarran, January 10, 1948; Patrick McCarran to John Dangberg, January 12, 1948; M.C.

Claar to John Dangberg, January 13, 1948; John Dangberg to W. P. Wing, January 17, 1948; John Dangberg to M. C. Claar, January 17, 1948; John Dangberg to A. E. Lawson, January 20, 1948; Gilbert F. Gower to John Dangberg, January 22, 1948; A. E. Lawson to John Dangberg, January 23, 1948; John Dangberg to Gilbert F. Gower, January 26, 1948; M. C. Claar to John Dangberg, February 4, 1948; W. P. Wing to John Dangberg, February 10, 1948, Box BSQAP 0114, Nevada Range Sheep Owners Association Papers, 1944–48, Jon Bilbao Basque Library.

31. Eva B. Adams to Grace M. Dangberg, February 10, 1948; John Dangberg to W. W. Whitaker, February 26, 1948; Harmon E. Hosier to the NRSOA, February 27, 1948; John Dangberg to Harmon E. Hosier, March 1, 1948; John Dangberg to Eva B. Adams, March 2, 1948; Harmon E. Hosier to John Dangberg, March 8, 1948; John Dangberg to Harmon E. Hosier, March 10, 1948, Box BSQAP 0114, Nevada Range Sheep Owners Association Papers, 1944–48, Jon Bilbao Basque Library; "Adopted Action," *Idaho Wool Growers Bulletin* 27, no. 7 (March 31, 1948): 6.
32. In 1944, Hatch Brothers of Wood Cross, Utah, contracted Karl V. King to import a group of Basque immigrants who arrived in 1945. Thereafter, Lawrence Jones from Malad, Idaho, requested him to import five Basque immigrant workers. Since both groups entered the United States on visitors' visas on six-month permits, they had to secure extensions every six months. In 1944, the Eastern Nevada Sheep Growers Association obtained a permit to import 150 Basque exiles from Mexico. All except thirteen of this group were subsequently deported or returned to Mexico. The ENSGA was required to obtain renewals of these immigrants' permits every six months, which placed a burden upon the association that they did not wish to assume. Thus, the ENSGA contracted King to undertake the responsibility of obtaining extensions for them, as well as for Jones and the Hatch Brothers. In 1945, it was reported from Spokane that there were ample sheepherders in the state of Washington. Therefore, the immigration authorities did not grant any extension to Jones, the Hatch Brothers, or the ENSGA until the surplus of herders available in the West was absorbed. King contacted the Yakima Sheep Company of Yakima, Washington. Since then, it became apparent the necessity of cooperation among all western woolgrowers desiring to import Basque herders on any future program. King assumed responsibility for obtaining extensions on all Basques, regardless of the employer or the place. King was successful in obtaining the necessary extensions, Report by Karl V. King, May 18, 1948, Box BSQAP 0114, Nevada Range Sheep Owners Association Papers, 1944–48, Jon Bilbao Basque Library.
33. Ibid.
34. These hearings were held in the following places: Pocatello, Idaho; Boise, Idaho; Spokane, Washington; Seattle, Washington; San Francisco, California; Sacramento, California; Reno, Nevada; Winnemucca, Nevada; Elko, Nevada; Ely, Nevada; and Salt Lake City, Utah. Ibid.
35. Ibid.
36. Ibid.
37. Ibid.
38. Ibid.

39. Ibid.
40. Ibid.
41. J.M. Jones to Berry N. Duff, W.P. Wing, and M.C. Claar, March 16, 1948, Box BSQAP 0114, Nevada Range Sheep Owners Association Papers, 1944–48, Jon Bilbao Basque Library.
42. Ibid.
43. Ibid.
44. Karl V. King to J.M. Jones, March 26, 1948; W.P. Wing to John Dangberg, March 30, 1948; John Dangberg to W.P. Wing, March 31, 1948; M.C. Claar to John Dangberg, April 2, 1948, Box BSQAP 0114, Nevada Range Sheep Owners Association Papers, 1944–48, Jon Bilbao Basque Library.
45. J.M. Jones to Berry N. Duff, W.P. Wing, and M.C. Claar, March 16, 1948; John Dangberg to W.P. Wing, March 31, 1948, Box BSQAP 0114, Nevada Range Sheep Owners Association Papers, 1944–48, Jon Bilbao Basque Library.
46. W.P. Wing to William Darsie, March 19, 1948, Box BSQAP 0114, Nevada Range Sheep Owners Association Papers, 1944–48, Jon Bilbao Basque Library.
47. Ibid.
48. J.M. Jones to Berry N. Duff, W.P. Wing, and M.C. Claar, March 16, 1948; Karl V. King to J.M. Jones, March 26, 1948, Box BSQAP 0114, Nevada Range Sheep Owners Association Papers, 1944–48, Jon Bilbao Basque Library.
49. Affidavit, John Dangberg and D.W. Park, May 12, 1948, State District Court, Douglas County Courthouse, Minden, Nevada, Box BSQAP 0114, Nevada Range Sheep Owners Association Papers, 1944–48, Jon Bilbao Basque Library.
50. W.P. Wing to the Board of Directors and Members of Advisory Committee, April 1, 1948; M.C. Claar to John Dangberg, April 2, 1948; M.C. Claar to Grace M. Dangberg, April 12, 1948; Patrick McCarran to Grace M. Dangberg, April 23, 1948; John Dangberg to the Nevada State Employment Service of Carson City, May 12, 1948; Affidavit, John Dangberg and D.W. Park, May 12, 1948, State District Court, Douglas County Courthouse, Minden, Nevada, Box BSQAP 0114, Nevada Range Sheep Owners Association Papers, 1944–48, Jon Bilbao Basque Library.
51. Karl V. King to the Secretaries of State Wool Growers Associations, May 18, 1948; Karl V. King to W.P. Wing, May 31, 1948, Box BSQAP 0114, Nevada Range Sheep Owners Association Papers, 1944–48, Jon Bilbao Basque Library.
52. H.R. 1572, Private Law 350, 80th Cong., 2d Sess., June 10, 1948 [Reported from Judiciary on December 16, 1947]; S. 158, Private Law 425, 80th Cong., 2d Sess., June 29, 1948 [Reported in Senate on May 17, 1948]; S. 298, Private Law 156, 80th Cong., 2d Sess., January 26, 1948 [Reported in Senate on June 24, 1947]; S. 1973, Passed Senate June 18, 1948, referred to Judiciary July 26, 1948 [Reported in Senate on June 15, 1948], *Calendars of the United States House of Representatives and History of Legislation, Eightieth Congress, Final Edition*, 82, 100, 225; S. 27, Private Law 13, 81st Cong., 1st Sess., April 14, 1949 [Reported in Senate on January 24, 1949], *Calendars of the United States House of Representatives and History of Legislation, Eighty-First Congress, Final Edition*, (Washington, D.C.:

Government Printing Office, 1951), 101; Report by Karl V. King, May 18, 1948; Karl V. King to the Secretaries of State Wool Growers Associations, May 18, 1948, Box BSQAP 0114, Nevada Range Sheep Owners Association Papers, 1944–48, Jon Bilbao Basque Library.

53. Alberto López Herce to German Baraibar, August 27, 1948; "Agreement," August 1948, Box C-110/15, "Pastores-Vascos," Central Archive of the Spanish Ministry of Labor, Madrid, Spain.
54. "Agreement," August 1948, Box C-110/15, "Pastores-Vascos," Spanish Ministry of Labor.
55. Ibid.; Alberto López Herce to German Baraibar, September 2, 1948; German Baraibar to Alberto López Herce, September 14, 1948; German Baraibar to the Spanish Minister of Foreign Affairs, September 14, 1948; "Contrato de Trabajo;" Spanish Subsecretary of P.A. (Foreign Affairs) to Subsecretary of the Ministry of Labor, October 15, 1948, Box C-110/15, "Pastores-Vascos," Central Archive of the Spanish Ministry of Labor.
56. "120,000 Sheep, Cattle Endangered by Storm," *Nevada State Journal*, January 21, 1949, 14; Beltran Paris (as told to William A. Douglass), *Beltran: Basque Sheepman of the American West* (Reno: University of Nevada Press, 1979), 169; "U.S. Airforce Ready to Fly Hay to Stock," *Nevada State Journal*, January 22, 1949, 14; "Airplanes Start Food Lift to Supply Starving Stock," *Reno Evening Gazette*, January 24, 1949; "Temperatures Drop Far Below Zero in Nevada as Cold Snap Continues," *Reno Evening Gazette*, January 25, 1949; *Salt Lake Tribune*, January 29, 1949, 6; "Air Force Reports Delivery of 1,178,945 Pounds of Hay to Starving Nevada Animals," *Nevada State Journal*, February 1, 1949, 2; "Operation Haylift Burns Gasoline to Feed Livestock," *Reno Evening Gazette*, February 5, 1949; "Elko Ranches, Towns Isolated by Storms," *Reno Evening Gazette*, February 16, 1949; "Dramatic Haylift Operation Ended," *Reno Evening Gazette*, February 17, 1949; "McCarran Asks $355,000 Fund to Fight Floods," *Reno Evening Gazette*, February 18, 1949; *Reno Evening Gazette*, December 31, 1949, 6; *Time* and *Life* magazines run stories on the winter of 1948–1949: "Death on the Range," *Time* 53, no. 6 (February 7, 1949): 13–14; "West Fights Worst Winter in History," *Life* 26, no. 7 (February 14, 1949): 19–23.
57. "Alien Basque Legislation," *Idaho Wool Growers Bulletin* 28, no. 6 (March 16, 1949): 2; S. 27, Private Law 13, 81st Cong., 1st Sess., April 14, 1949 [Reported in Senate on January 24, 1949], *Calendars of the United States House of Representatives and History of Legislation, Eighty-First Congress, Final Edition*, 101, 238, 306.
58. Thomas J. Carty, "White House Outreach to Catholics," in *Catholics and Politics: The Dynamic Tension between Faith and Power*, eds. Kristen E. Heyer, Mark J. Rozell, and Michael A. Genovese (Washington, D.C.: Georgetown University Press, 2008), 177–80; to learn more about the political role of the American Catholic Church that emerged during the New Deal and how it consolidated its power during the early part of the Cold War, see: Donald F. Crosby, SJ, *God, Church, and Flag: Senator Joseph R. McCarthy and the Catholic Church, 1950–1957* (Chapel Hill: University of North Carolina Press, 1978); Thomas M. Moriarty,

"The Catholic Lobby: The Periphery Dominated Center, Public Opinion and American Foreign Policy, 1932–1962" (PhD diss., University of Massachusetts, Amherst, 1996); David B. Woolen and Richard G. Kurial, eds., *FDR, the Vatican, and the Roman Catholic Church in America, 1933–1945* (New York: Palgrave Macmillan, 2003); Steve Rosswurm, *The FBI and the Catholic Church, 1935–1962* (Amherst: University of Massachusetts Press, 2009).

59. "McCarran Threatens Budget Cut," *New York Times*, May 6, 1949, 11; *Public Papers of the Presidents of the United States: Harry S. Truman, 1949* (Washington, D.C.: Government Printing Office, 1964), 380; "President Rejects Franco Loan Plan," *New York Times*, July 15, 1949, 6.
60. "M'Carran Fears DP Burden Here," *New York Times*, September 15, 1949, 3; "Franco Said to Eye Pact for U.S. Aid," *New York Times*, November 1, 1949, 19; "M'Carran on His Own on Trip, Truman Says," *New York Times*, September 16, 1949, 3.
61. "Franco Said to Eye Pact for U.S. Aid," *New York Times*, November 1, 1949, 19; "El jefe del estado recibió ayer a Míster Mac Carran," *ABC*, November 3, 1949, 3.
62. "El jefe del estado recibió ayer a Míster Mac Carran," *ABC*, November 3, 1949, 1, 3, 17; on the early American aid to Franco's Spain, see: Juan Dura, "U.S. Policy toward Dictatorship and Democracy in Spain, 1936–1953: A Case Study in Realities of Policy Formation" (PhD diss., University of California, Berkeley, 1979); Enrique Fanjul, "El papel de la ayuda americana en la economía española, 1951–1957," *Información Comercial Española* no. 577 (September 1981): 159–65; Ángel Viñas, *Guerra, dinero, dictadura: Ayuda fascista y autarquía en la España de Franco* (Barcelona: Crítica, 1984); J. Carter Murphy and R. Richard Rubottom, *Spain and the United States since World War II* (New York: Praeger, 1984); Fernando Guirao, *Spain and the Reconstruction of Western Europe, 1945–1957: Challenge and Response* (London: Macmillan, 1998); Calvo González, "¡Bienvenido, Míster Marshall!;" Lorenzo Delgado Gómez-Escalonilla, "¿El 'amigo americano'? España y Estados Unidos durante el franquismo," *Studia Histórica* 21 (2003): 231–76.
63. "McCarran Believes Franco Democratic," *Austin American*, November 3, 1949, 23; "Franco Trying to Win Friendship," *Reno Evening Gazette*, November 3, 1949; "Senator Wants Spain in Pact," *Reno Evening Gazette*, November 4, 1949, 11; "Los senadores norteamericanos Dewey y Mac Carran conversando," *ABC*, November 4, 1949, 5; "Meddling in Foreign Affairs," *Reno Evening Gazette*, November 23, 1949, 4.
64. Ybarra, *Washington Gone Crazy*, 466–84.
65. The article was reprinted in other newspapers nationwide, such as: *Winona Republican Herald*, November 22, 1949, 8; *Harrisburg Daily Register*, November 23, 1949, 2; *Ogden Standard-Examiner*, November 23, 1949, 4; *Brownsville Herald*, November 23, 1949, 4; *Lewiston Daily Sun*, November 23, 1949, 4; *Idaho State Journal*, November 24, 1949, 4; *Camden News*, November 27, 1949; *Alto Herald*, December 22, 1949; *Moulton Eagle*, December 23, 1949.
66. Robert S. Allen, "Looking 'Em Over," *Abilene*, September 23, 1949, 6.
67. The *Reno Evening Gazette* followed closely McCarran's visit to Franco: "McCarran Seeks To Support Case," *Reno Evening Gazette*, November 2, 1949;

"Senator Chats with Franco," *Reno Evening Gazette*, November 3, 1949; "Franco Trying to Win Friendship," *Reno Evening Gazette*, November 3, 1949; "Senator Wants Spain in Pact," *Reno Evening Gazette*, November 4, 1949; "U.S. Senator Visits Franco," *Reno Evening Gazette*, November 11, 1949; "Meddling in Foreign Affairs," *Reno Evening Gazette*, November 23, 1949, 4.

68. "Franco Trying to Win Friendship," *Reno Evening Gazette*, November 3, 1949; "U.S. Senator Visits Franco," *Reno Evening Gazette*, November 11, 1949; "House Member Off for Shipyard Tour," *New York Times*, September 8, 1949, 32; "U.S. Group to See Franco," *New York Times*, September 14, 1949, 15; "U.S. Admiral Sees Franco," *New York Times*, September 6, 1949, 12.
69. Powaski, *Cold War*, 76, 80–81.
70. Carlos Manzanares, "Informe para el Sr. Ministro," February 11, 1949, Box C-110/15, Central Archive of the Spanish Ministry of Labor.
71. Ibid.; Subsecretary of the Spanish Ministry of Labor to the Subsecretary of the Spanish Ministry of Governance, March 1, 1949, Box C-110/15, Central Archive of the Spanish Ministry of Labor.
72. David M. Oshinsky, *A Conspiracy So Immense: The World of Joe McCarthy* (New York: Free Press, 1983), 108–9, 168; Powaski, *Cold War*, 86–87, 89–92.
73. "Senate Votes Spain Loan of $100,000,000 to Put her in E.R.P." *New York Times*, August 2, 1950, 1; "Truman Criticizes Huge Loan to Spain," *New York Times*, August 4, 1950, 6; "Spain Is Jubilant on U.S. Loan Action," *New York Times*, August 3, 1950, 18; "El Proyectado Crédito Norteamericano a España," *ABC*, August 5, 1950, 9; Viñas, *Los pactos secretos de Franco con Estados Unidos*, 61; historian Ellen Schrecker uses the term "professional anti-Communists" referring those individuals who worked consciously in various areas of the federal government on behalf of the anticommunist cause. *Many Are the Crimes*, 43–45.
74. "Senate for Basque Residence," *New York Times*, August 24, 1950, 10; Edwards, *Pat McCarran*, 193–94.
75. The McCarran-Walter Act of 1952 has stirred up great academic interest among scholars. See, for example: Richard Harrison Robbins, "The Immigration Act of 1952: A Case Study in Political Sociology" (PhD diss., University of Illinois, 1959); Erwin Alan Jaffe, "Passage of the McCarran-Walter Act: The Reiteration of American Immigration Policy" (PhD diss., Rutgers University, 1962); Marius Albert Dimmitt Sr., "The Enactment of the McCarran-Walter Act of 1952" (PhD diss., University of Kansas, 1970); there are many other studies that pay special attention to the Immigration and Nationality Act. See: Michael Gill Davis, "The Cold War, Refugees, and United States Immigration Policy, 1952–1965" (PhD diss., Vanderbilt University, 1996); Cheryl Shanks, *Immigration and the Politics of American Sovereignty, 1890–1990* (Ann Arbor: University of Michigan Press, 2001); Carl Joseph Bon Tempo, "Americans at the Gate: The Politics of American Refugee Policy, 1952–1980" (PhD diss., University of Virginia, 2004); Ngai, *Impossible Subjects*; Danielle Battisti, "Relatives, Refugees, and Reform: Italian Americans and Italian Immigration During the Cold War, 1945–1965" (PhD diss., State University of New York at Buffalo, 2010).

76. Ngai, *Impossible Subjects*, 234–238; Oscar Handlin, "The Immigration Fight Has Just Begun," *Commentary* 14 (July 1952): 4.
77. John P. Bidegaray and Philip Erro to the CRA Members, October 8, 1951; "Important Bulletin for Prospective Sponsors in Proposed Sheepherder Importation Program!" Box BSQAP 0112, Western Range Association Papers, Jon Bilbao Basque Library; "Bidegaray Resigns California Range Association Office," *National Wool Grower* 48 (1958): 24.
78. John P. Bidegaray and Philip Erro to the Members of the CRA, October 8, 1951, Box BSQAP 0112, Western Range Association Papers; John Dangberg to Robert Franklin, October 3, 1951, Box BSQAP 0116, Nevada Range Sheep Owners Association Papers, 1944–48, Jon Bilbao Basque Library.
79. S. 2549, Public Law 307, 82nd Cong., 2nd Sess., April 9, 1952; Special Bulletin from John P. Bidegaray to the Members of the CRA, January 30, 1952; John P. Bidegaray to the Members of the CRA, February 25, 1952, Box BSQAP 0112, Western Range Association Papers; "Bidegaray & Franklin Return from Washington," *California Livestock News* 28, no. 7 (February 12, 1952): 5.
80. "Sheepherder Bill Awaits President's Signature," *National Wool Grower* 42, no. 4 (April 1952): 7.
81. John Dangberg to Pete Arraras, April 4, 1952; John Bidegaray to all Wool Growers in Nevada, May 9, 1952, Box BSQAP 0119, Nevada Range Sheep Owners Association Papers, 1944–48, Jon Bilbao Basque Library; John Bidegaray to the Members of CRA, April 12, 1952; John Bidegaray to the Members of CRA, May 14, 1952; Robert Franklin to all sponsors, August 11, 1952; Report by John Bidegaray and Robert Franklin, August 20, 1952, Box BSQAP 0112, Western Range Association Papers; "An Act to Provide Relief for the Sheep-Raising Industry by Making Special Quota Immigration Visas Available to Certain Alien Sheepherders," Public Law 307, *United States Statutes at Large*, vol. 66, 82nd Cong. (Washington, D.C.: Government Printing Office, 1953), *vii*, 50–51; *Reno Evening Gazette*, July 21, 1952; "Preliminary Interviews Begin In Bilbao, Range Assn. Officers Report," *California Livestock News* 27, no. 34 (August 19, 1952); "Spanish Visas Delayed For Basques Needed As Sheepherders In U.S.," *California Livestock News* 27, no. 35 (August 26, 1952); "Franklin Tells 19 Sponsors Of Herder Processing; Two Herders Are Ineligible," *California Livestock News* 27, no. 36 (September 2, 1952): 3.
82. John Bidegaray to the Members of the CRA, February 5, 1952, Box BSQAP 0112, Western Range Association Papers; "McCarran-Walter Immigration Bill Becomes Law as Congress Overrides President Truman's Veto," *California Livestock News* 27, no. 27 (July 1, 1952): 1, 8.
83. Peter Edson, "McCarran Act Backfires on Its Sheepish Author," *Times-News*, March 13, 1951, 2; the article was reprinted in other newspapers nationwide, such as: *Coshocton Tribune*, March 10, 1951, 5; *Kalispell Daily Inter Lake*, March 13, 1951; *Mexia Daily News*, March 13, 1951, 4; *Bonham Daily Favorite*, March 14, 1951, 2.
84. The Spanish *Falange* was the only legal party in the hands of the Francisco Franco during his dictatorship, operating from 1937 until its dissolution in 1977. Works

that emphasize the nature of the *Falange* and the domestic political evolution of Franco's Spain include: Stanley G. Payne, *Falange: A History of Spanish Fascism* (Stanford, Calif.: Stanford University Press, 1961); Javier Tusell, *Franco y los católicos: La política interior española entre 1945 y 1957* (Madrid: Alianza Editorial, 1984); Joan Maria Thomàs, *La Falange de Franco: Fascismo y fascistización en el régimen franquista (1937–1945)* (Barcelona: Plaza & Janés Editores, 2001); and more recently, Ferran Gallego, *El evangelio fascista: La formación de la cultura política del franquismo (1930–1950)* (Barcelona: Crítica, 2014); for a general sociopolitical overview of the Franco's dictatorship, see: Stanley G. Payne, *The Franco Regime, 1936–1975* (Madison: University of Wisconsin Press, 1987); Moradiellos, *La España de Franco (1939–1975)*.

85. Peter Edson, "McCarran Act Backfires on Its Sheepish Author," *Times-News*, March 13, 1951, 2; "Basque Herders May Be Brought to United States," *Evening Courier*, January 29, 1951, 3.
86. "Basque Shepherd: U.S. Immigrant Works Amid the Alien Lambs," *Life* 32, no. 19 (May 12, 1952): 125, 127.
87. Camille M. Cianfarra, "Basques Are Best: The World's Finest Shepherds are Coming Here to Take Care for our Native Western Flocks," *New York Times*, November 23, 1952, 78.
88. Ibid.; later, in 1955, Vivian Rowe described the Basque people as follows: "They are a race of the open-air: hard-working, frugal, fond of manly sports, of singing and of dancing." Vivian Rowe, *The Basque Country* (London: Putman, 1955), 11.
89. "County Fair Plans on Special Day to Honor McCarran," *Nevada State Journal*, July 11, 1953, 2; "Sen. McCarran to Be Honored at County Fair," *Nevada State Journal*, August 7, 1953, 8; "100 Nevadans to Help Honor Sen. McCarran," *Nevada State Journal*, September 3, 1953, 7; "Nation's Great Song Writers Presenting Show Here on 19th to Honor Sen. Pat McCarran," *Nevada State Journal*, September 6, 1953, 14; "Gift Is Prepared for M'Carran Day," *Nevada State Journal*, September 9, 1953, 2; "Elaborate Plans Are Being Made for Honoring Senator McCarran Here on Saturday," *Nevada State Journal*, September 16, 1953, 20; "This Is McCarran Day; Many Will Take Part," *Nevada State Journal*, September 19, 1953, 3; "Thousands Jam Fairgrounds to Honor Senator," *Nevada State Journal*, September 20, 1953, 14; "350 Attend Luncheon Honoring Sen. McCarran," *Nevada State Journal*, September 20, 1953, 3.
90. "El Embajador de España en Washington ha impuesto al senador Mac Carran la Gran Cruz de Isabel la Catolica," *ABC*, August 26, 1953, 14; "Condecoración española al senador norteamericano Mac Carran," *ABC*, September 1, 1953, 7; "Condecoración española al senador norteamericano Mac Carran," *ABC*, September 2, 1953, 1; "Franco Decorates McCarran," *New York Times*, October 3, 1953, 8; Lowi, *American Civil-Military Decisions*, 697.
91. Moradiellos, *La España de Franco (1939–1975)*, 99–102; Viñas, *En las garras del águila*, 512–14; John Dangberg to Peter Arraras, February 12, 1952; Peter Arraras to John Dangberg, February 8, 1952; John Dangberg to Peter Arraras, April 4, 1952; John Dangberg to Peter Arraras, June 6, 1952; John Dangberg to Peter Arraras,

June 23, 1952; Peter Arraras to John Dangberg, July 1, 1952; Peter Arraras to John Dangberg, August 6, 1952; Peter Arraras to John Dangberg, September 4, 1952; Peter Arraras to John Dangberg, September 27, 1952; Peter Arraras to John Dangberg, October 29, 1952; Peter Arraras to John Dangberg, October 30, 1952; Peter Arraras to John Dangberg, November 18, 1952; Peter Arraras to John Dangberg, December 30, 1952; Peter Arraras to John Dangberg, January 16, 1953; Peter Arraras to John Dangberg, January 28, 1953; Peter Arraras to John Dangberg, February 20, 1953; Peter Arraras to John Dangberg, March 11, 1953; Peter Arraras to John Dangberg, April 2, 1953, Box BSQAP 0119, Nevada Range Sheep Owners Association Papers, 1944–48, Jon Bilbao Basque Library; Pedro Salvador to the Spanish Bureau of Consular Affairs, October 17, 1955, Box C-110/15, Central Archive of the Spanish Ministry of Labor; *ABC*, January 26, 1955; Note from the Spanish Ministry of Labor to Juan Francisco de Cárdenas, December 7, 1954; Director General of Spanish Foreign Affairs to the Director General of the Spanish Ministry of Labor, January 29, 1955; Eduardo Propper de Callejón to the Spanish Foreign Ministry, August 10, 1955; Director General of Spanish Foreign Affairs to the Director General of the Spanish Ministry of Labor, August 18, 1955; Pedro Salvador to the Spanish Bureau of Consular Affairs, October 17, 1955; Director General of Spanish Foreign Affairs to the Director General of the Spanish Ministry of Labor, November 3, 1955, Box C-110/15, Central Archive of the Spanish Ministry of Labor.

92. "That, for a period of one year after the effective date of this Act, in any case in which the Attorney General, under the authority of section 204 of the Immigration and Nationality Act, grants permission for the importation of a skilled alien sheepherder into the United States and the investigation of the application for such importation discloses that: 1) the employment offered such skilled alien sheepherder is permanent; and, 2) no immigration quota number of the country of which such skilled alien sheepherder is a national is then available, a special nonquota immigration visa may be issued to such skilled alien sheepherder as provided in this Act: Provided, That such skilled alien sheepherder is otherwise admissible into the United States for permanent residence." S. 2862, 83d Cong., 2d Sess., February 1, 1954, Box BSQAP 0112, Western Range Association Papers.
93. S. 2862, 83d Cong., 2d Sess., February 1, 1954; Dwight D. Eisenhower, "Statement by the President Upon Signing Bill Making Available Special Nonquota Immigrant Visas to Skilled Alien Sheepherders," September 3, 1954; S. 2862, Public Law 770, 83d Cong. [68 Stat. 1145], Box BSQAP 0112, Western Range Association Papers.
94. "Senator M'Carran Is Dead in Nevada," *New York Times*, September 29, 1954, 1; "El Senador Mac Carran ha muerto repentinamente en Hawthorne (Nevada), después de pronunciar un discurso," *ABC*, September 30, 1954, 11; Douglas T. Miller and Marion Nowak, *The Fifties: The Way We Really Were* (Garden City, New York: Doubleday and Company, 1977), 5–18, 36–37; Vincent J. Cannato, *American Passage: The History of Ellis Island* (New York: Harper Collins, 2009), 14; Ratner et al., *Evolution of the American Economy*, 458–61.

Acknowledgments

The present book would not have been possible without the support and assistance of countless individuals. I would like to begin by thanking the most important people, my family—my parents, Ione Idarraga and Eduardo Saitua, and my sister, Leire—for their continuous encouragement and support along the way. To the rest of my large extended family (you know who you are), I extend a loving thank you. I would also like to thank my long time friends from my hometown Algorta for understanding my frequent absence during the writing process over the last years.

This book is based on research I conducted during my graduate work at University of Nevada, Reno. I would first like to extend my heartfelt and sincere thanks to William D. Rowley for his excellent advice on my research. His great knowledge on the topics addressed in this book and guidance has been absolutely invaluable for my work. Also, I am indebted to Coro Rubio Pobes who has encouraged me in my academic career since my undergraduate years. She guided me with critical insight and provided detailed advice about countless methodological issues on my work. Her assistance and advice have always been of enormous help. I am grateful also to Xabier Irujo, Kevin D. Hatfield, Paul Starrs, and Joseba Zulaika for their insightful comments and suggestions.

Numerous librarians, archivists, and manuscript curators patiently assisted me with my research. To the following professionals, I owe special gratitude: Iñaki Arrieta Baro, Jon Bilbao Basque Library, University of Nevada, Reno; Michael P. Maher, Nevada Historical Society; Donnelyn Curtis, Special Collections, University of Nevada, Reno; Toni L. Mendive, Northeastern Nevada Museum; Patty Miller, Basque Museum and Cultural Center; Ree Taylor, Eureka Sentinel Museum; Mike Smales, Elko County Courthouse; Tami Rae Spero, Humboldt County Courthouse; Sadie Sullivan, Lander County Courthouse; Susan Lujan, White Pine County Courthouse; Luis Casado, General Records of the Spanish Department of Labor.

I have been greatly helped in writing the present book by several scholars and colleagues. My particular debt here is to Aurora Bosch who from the start was a constant source of help, advice, and encouragement. I would also like to thank Devra Weber for reading and commenting on parts of my manuscript and providing constructive suggestions. In addition, I would like to thank the following people who made valuable contributions in various ways: Ryan Powell, Sandra Ott, Dave Tilford, Donald Drake, Lyle McNeal, Michelle Alzola, Anita Anacabe, and Denys Koyle. I am also extremely grateful for the help of other friends, staff members, and students from my time as a graduate student at UNR: Iñaki and Gina Asueta (and the rest of the family), Kate Camino, Tania Arriaga, Nagore Sedano, Amaia Iraizoz, Ziortza Gandarias, Patricia A. DeBunch, and James H. Bunch.

I am immensely indebted to Justin Race, my editor at the University of Nevada Press, for his guidance and encouragement throughout this entire process. Thank you also to the two anonymous reviewers of my manuscript for their insightful suggestions that have been incorporated into this book.

Some parts of the present book include revised versions and parts of some of my published articles. I want to acknowledge the permission granted by the publishers to reproduce them. Chapter 2 is an extended and revised version of the article, "Becoming Herders: Basque Immigration, Labor, and Settlement in Nevada, 1880–1910," *Montana, The Magazine of Western History* 66, no. 4 (Winter 2016), 58–70. The introduction, chapter 2, and chapter 4 contain revised parts of the article, "*The Best Sheepherder*: The Racial Stereotype of Basque Immigrants in the American West Between the End of the Nineteenth and the Beginning of the Twentieth Centuries," *Historia Contemporánea* 1, no. 56 (2018), 81–119. And finally, chapter 4 contains revised parts of the article, "Resistance to the 'Hoofed Locusts': Key Pittman, Basque Immigrants, and the Sheep Industry of Nevada, 1910–1920," *Nevada Historical Society Quarterly* 60, no. 1–4 (2017), 148–172.

Bibliography

Collections of Papers and Documents

Aguirre, Angela. Papers. Aguirre Family Archive, Elko, Nevada.

Barinaga-Oxarango Family Archive, Boise, Idaho.

Basque Sheepherders Papers. Central Archive of the Spanish Ministry of Labor. Madrid, Spain.

Civil and Criminal Records. Elko County Courthouse. Elko, Nevada.

Civil and Criminal Records. Humboldt County Courthouse. Winnemucca, Nevada.

Civil and Criminal Records. Lander County Courthouse, Battle Mountain, Nevada.

Civil and Criminal Records. Washoe County Courthouse, Reno, Nevada.

Civil and Criminal Records. White Pine County Courthouse, Ely, Nevada.

County Deeds. Elko County Courthouse. Elko, Nevada.

Gisasola, Emeterio. Papers, Gisasola-Rementeria Family Archive, Ibarrangelu, Bizkaia, Spain.

Marriage Records. Elko County Courthouse. Elko, Nevada.

McCarran, Patrick. Papers. Nevada Historical Society, Reno, Nevada.

Naturalization Records. Elko County Courthouse. Elko, Nevada.

Nevada Range Sheep Owners Association Papers. Mathewson-IGT Knowledge Center. Jon Bilbao Basque Library, University of Nevada, Reno.

New York Passenger Arrival Records, 1820–1957, National Archives and Records Administration, New York City, New York.

Pittman, Key. Papers. Library of Congress, Washington, D.C.

Pittman, Key. Papers. Nevada Historical Society, Reno, Nevada.

Ranching Papers. Northeastern Nevada Museum, Elko, Nevada.

Records of the U.S. Forest Service. National Archives and Records Administration, College Park, Maryland.

Records of the U.S. Forest Service. National Archives and Records Administration. National Archives at San Francisco, California.

Rothstein, Arthur. Photographic Collection. Library of Congress, Washington, D.C.

Uberuaga Hormaechea, Juanita. Papers. Basque Museum and Cultural Center, Boise, Idaho.

U.S. Department of the Interior. Bureau of Land Management. General Land Office Records, Washington, D.C.

Vertical Files. Manuscript Collection. Eureka Sentinel Museum, Eureka, Nevada.

Vertical Files. Nevada Historical Society, Reno, Nevada.

Western Range Association Papers. Mathewson-IGT Knowledge Center. Jon Bilbao Basque Library, University of Nevada, Reno.

Williams Family Archive, Fallon, Nevada.

Courthouse Legal Cases and Disputes

C.H. Duborg et al. v. *Juan Bidart*, case no. 2069, 1901, Lander County Courthouse, Battle Mountain, Nevada.

The State of Nevada v. *Domingo Echeverria*, case no. 319, 1951, Humboldt County Courthouse, Winnemucca, Nevada.

Oral History Transcripts

Aguirre, Angela. Oral history by Bill Price, Northeastern Nevada Museum, Elko, Nevada, August 27, 1993.

Artecheverria Gabicagogeascoa, Mari. Oral history by Joseba Chertudi, Basque Museum and Cultural Center, Boise, Idaho, January 16, 1976.

Astorquia Osborne, Lucy. Oral history by Daniel Chertudi and Julian Achabal, Basque Museum and Cultural Center, Boise, Idaho, June 26, 2002.

Barinaga, Pedro "Pete." Oral history by Joseba Chertudi, Basque Museum and Cultural Center, Boise, Idaho, August 20, 1975.

Lugea Arregui, Elena and Marianne Lugea Goicoechea. Oral history by Begoña Pecharroman, Northeastern Nevada Museum, Elko, Nevada, June 2, 1998.

Olechea, Juanita. Oral history by Daniel Chertudi, Basque Museum and Cultural Center, Boise, Idaho, September 4, 2001.

Yraguen Echanis, Timotea. Oral history by Patty A. Miller, Basque Museum and Cultural Center, Boise, Idaho, March 16, 2002.

Yribar, Juanita. Oral history by Mikel Chertudi, Basque Museum and Cultural Center, Boise, Idaho, September 14, 2001.

Yturri, Justa. Oral history by Joseba Chertudi and Begoña Pecharroman, Basque Museum and Cultural Center, Boise, Idaho, August 2, 1975.

Yturri, Reme. Oral history by Patty A. Miller, Basque Museum and Cultural Center, Boise, Idaho, March 16, 2002.

Theses and Dissertations

Araujo, Frank P. "Basque Cultural Ecology and Echinococcosis in California." PhD diss., University of California, Davis, 1974.

Arrizabalaga, Marie-Pierre. "A Statistical Study of Basque Immigration into California, Nevada, Idaho, and Wyoming between 1900 and 1910." Master's thesis, University of Nevada, Reno, 1986.

Baker, Sarah C. "Basque American Folklore in Eastern Oregon." Master's thesis, University of California, Berkeley, 1972.

Battisti, Danielle. "Relatives, Refugees, and Reform: Italian Americans and Italian Immigration during the Cold War, 1945–1965." PhD diss., State University of New York at Buffalo, 2010.

Bon Tempo, Carl Joseph. "Americans at the Gate: The Politics of American Refugee Policy, 1952–1980." PhD diss., University of Virginia, 2004.
Bouesnard, Mademoiselle. "Basque Emigration to California and Nevada since 1960." Master's thesis, Université de Pau et des Pays de l'Adour, 1976.
Castelli, Joseph R. "Basques in the Western United States: A Functional Approach to Determination of Cultural Presence in the Geographic Landscape." PhD diss., University of Colorado, 1970.
Davis, Michael Gill. "The Cold War, Refugees, and United States Immigration Policy, 1952–1965." PhD diss., Vanderbilt University, 1996.
Decroos, Jean F. "The Long Journey: Assimilation and Ethnicity Maintenance among Urban Basques in Northern California." PhD diss., University of Oregon, 1979.
Dimmitt, Marius Albert, Sr. "The Enactment of the McCarran-Walter Act of 1952." PhD diss., University of Kansas, 1970.
Dunn, Louise B. "The Salt Lake City Basque Community: Atypical in the American West." Master's thesis, University of Utah, 1972.
Dura, Juan. "U.S. Policy toward Dictatorship and Democracy in Spain, 1936–1953: A Case Study in Realities of Policy Formation." PhD. diss, University of California, Berkeley, 1979.
Eagle, Sonia J. "Work and Play among the Basques of Southern California." PhD diss., Purdue University, 1979.
Edlefsen, John B. "A Sociological Study of the Basques of Southwest Idaho." PhD diss., State College of Washington, 1948.
Gaiser, Joseph H. "The Basques of the Jordan Valley Area: A Study in Social Processes and Social Change." PhD diss., University of Southern California, 1944.
Gómez-Ibáñez, Daniel Alexander. "The Rise and Decline of Transhumance in the United States." Master's thesis, University of Wisconsin, 1967.
Harkness, Ione B. "Certain Community Settlements in Oregon." Master's thesis, University of Southern California, 1925.
Hatfield, Kevin D. "'We Were Not Tramp Sheepmen': Resistance and Identity in the Oregon Basque Community, Accustomed Range Rights, and the Taylor Grazing Act, 1890–1955." PhD diss., University of Oregon, 2003.
Jaffe, Erwin Alan. "Passage of the McCarran-Walter Act: The Reiteration of American Immigration Policy." PhD diss., Rutgers University, 1962.
Kelly, James P. "The Settlement of Basques in the American West." Senior paper, Harvard University, May 1967.
Lane, Richard H. "The Cultural Ecology of Sheep Nomadism: Northeastern Nevada, 1870–1972." PhD diss., Yale University, 1974.
McCall, Grant E. "Basque-Americans and a Sequential Theory of Migration and Adaptation." Master's thesis, San Francisco State College, 1968.
McCullough, Flavia M. "The Basques in the Northwest." Master's thesis, University of Portland, 1945.
Moriarty, Thomas M. "The Catholic Lobby: The Periphery Dominated Center, Public Opinion and American Foreign Policy, 1932–1962." PhD diss., University of Massachusetts, Amherst, 1996.

Pagliarulo, Carol M. "Basques in Stockton: A Study of Assimilation." Master's thesis, College of the Pacific, 1948.
Pickett, Evelyne L. "Women in the Empty Quarter: A Study of Changes and Challenges as Related to Women's Experiences in the Nevada Ranching Industry from the Mid-Nineteenth Century to the Late Twentieth." Master's thesis, University of Nevada, Reno, 1988.
Pittman, Von V. Jr., "Senator Patrick A. McCarran and the Politics of Containment." PhD diss., University of Georgia, 1979.
Rader, Harold G. "The McCarran Doctrine: An Abrogation of Federal Sovereign Immunity through the McCarran Adjudication Amendment." Master's thesis, University of Nevada, Reno, 1993.
Robbins, Richard Harrison. "The Immigration Act of 1952: A Case Study in Political Sociology." PhD diss., University of Illinois, 1959.
Ruiz, Allura N. "The Basques-Sheepmen of the West." Master's thesis, University of Nevada, 1964.
Sather, Clifford A. "Marriage Patterns among the Basques of Shoshone, Idaho." Master's thesis, Reed College, 1961.
Stafford, John A. "Basque Ethnohistory in Kern County, California: 1872–1934 A.D." Master's thesis, Sacramento State College, 1971.
Wilcox, Mary S. "A Historical Study of the Basque Race with Special Reference to the United States." Master's thesis, University of Utah, 1939.

Unpublished Manuscripts

Etcheverry, Mary Jane. "Autobiography." Eureka Sentinel Museum, Eureka, Nevada.
Sack, Ivan. "History of Toiyabe National Forest." 1965. Mathewson-IGT Knowledge Center. Special Collections, University of Nevada, Reno, Nevada.
Wines, Lourinda R. "The Long Journey." 1973. Mathewson-IGT Knowledge Center. Special Collections, University of Nevada, Reno, Nevada.

Published Material

Government Publications

Calendars of the United States House of Representatives and History of Legislation, Seventy-Seventh Congress, Final Edition. Washington, D.C.: Government Printing Office, 1942.
Calendars of the United States House of Representatives and History of Legislation, Seventy-Eighth Congress, Final Edition. Washington, D.C.: Government Printing Office, 1944.
Calendars of the United States House of Representatives and History of Legislation, Seventy-Ninth Congress, Final Edition. Washington, D.C.: Government Printing Office, 1946.
Calendars of the United States House of Representatives and History of Legislation, Eightieth Congress, Final Edition. Washington, D.C.: Government Printing Office, 1948.
Calendars of the United States House of Representatives and History of Legislation,

Eighty-First Congress, Final Edition. Washington, D.C.: Government Printing Office, 1951.

Clawson, Marion. "Range Lands of Northeastern Nevada: Their Proper and Profitable Use (A Progress Report)," Bureau of Agricultural Economics, USDA. Washington, D.C.: Government Printing Office, 1938.

Gates, Paul W., and Robert W. Swenson. *History of Public Land Law Development*. Washington, D.C.: Government Printing Office for Public Land Law Review Commission, 1968.

Hochmuth, H. R., Earl R. Franklin, and Marion Clawson. *Sheep Migration in the Intermountain Region*, USDA Circ. 624. Washington, D.C.: Government Printing Office, Jan. 1942.

The Miscellaneous Documents of the House of Representatives for the First Session of the Fifty-Second Congress, 1891–92. Vol. 50, Part 10. Washington, D.C.: Government Printing Office, 1896.

Public Papers of the Presidents of the United States: Harry S. Truman, 1949. Washington, D.C.: Government Printing Office, 1964.

Report in Regard to the Range and Ranch Cattle Business of the United States, by Joseph Nimmo Jr. Washington, D.C.: Government Printing Office, 1885.

Report of the Public Lands Commission, With Appendix, 58th Cong., 3d Sess., Doc. no. 189. Washington, D.C.: Government Printing Office, 1905.

Statutes of the United States of America, Passed at the Second Session of the Sixty-Fourth Congress, 1916–1917, Part I: Public Acts and Resolutions. Washington, D.C.: Government Printing Office, 1917.

Twelfth Annual Report of the Secretary of Labor. For the Fiscal Year Ended June 1924. Washington, D.C.: Government Printing Office, 1924.

Twelfth Census of the United States, Taken in the Year 1900, Agriculture, Part 1, Farms, Live Stock, and Animal Products. Washington, D.C.: United States Census Office, 1902.

Twelfth Census of the United States, Taken in the Year 1900, William R. Merriam, Dir. Population, Part 2. Washington, D.C.: United States Census Office, 1902.

United States Department of Agriculture, *1905 "Use Book": The Use of the National Forest Reserves; Regulations and Instructions*. Washington, D.C.: Government Printing Office, 1905.

United States Statutes at Large, vol. 66, 82nd Cong. Washington, D.C.: Government Printing Office, 1953.

Wooton, Elmer O. *The Public Domain of Nevada and Factors Affecting Its Use*, USDA Tech. Bulletin no. 301. Washington, D.C.: Government Printing Office, Apr. 1932.

Nevada State Documents

Appendix to the Journals of Senate and Assembly of the Tenth Session of the Legislature of the State of Nevada. Carson City: State Printing Office, 1881.

Biennial Report of the Surveyor-General and State Land Register, State of Nevada, for 1899–1900. Carson City: State Printing Office, 1901.

Biennial Report of the Surveyor-General and State Land Register, 1903–1904. Carson City: State Printing Office, 1905.

Biennial Report of the Surveyor-General and State Land Register of the State of Nevada, 1909–1910. Carson City: State Printing Office, 1911.

Flemming, Charles E. and C.A. Brennen. *Studies of the Range Sheep Business in Nevada,* Bulletin no. 145. Reno: University of Nevada, Reno, Aug. 1937.

Heller, Dean, Secretary of State. *Political History of Nevada (Eleventh Edition).* Carson City: State Printing Office, 2006.

Report of the Surveyor General and State Land Register of the State of Nevada, For the Years 1881 and 1882. Carson City: State Printing Office, 1883.

Report of the U.S. Surveyor General of Nevada for the fiscal year ending June 30th, 1885. Carson City: State Printing Office, 1885.

Second Biennial Report of the Commissioner of Labor of the State of Nevada, 1917–1918. Carson City: State Printing Office, 1919.

Shamberger, Hugh A. "Evolution of Nevada's Water Laws, as Related to the Development and Evaluation of the State's Water Resources from 1866 to about 1960." *Water-Resources, Bulletin 46.* Carson City: Nevada Division of Water Resources, 1991.

Statutes of the State of Nevada, Passed at the First Session of the Legislature, 1864–5. Carson City: State Printer, 1865.

Statutes of the State of Nevada Passed at the Sixth Session of the Legislature, 1873. Carson City: State Printer, 1873.

Statutes of the State of Nevada Passed at the Fourteenth Session of the Legislature, 1889. Carson City: State Printing Office, 1889.

Statutes of the State of Nevada, Passed at the Sixteenth Session of the Legislature, 1893. Carson City: State Printing Office, 1893.

Statutes of the State of Nevada, Passed at the Twenty-First Session of the Legislature 1903. Carson City: State Printing Office, 1903.

Statutes of the State of Nevada, Thirty-First Session of the Legislature 1923. Carson City: State Printing Office, 1923.

The Journal of the Assembly of the Seventeenth Session of the Legislature of the State of Nevada, 1895. Carson City: State Printing Office, 1895.

The Journal of the Assembly of the Eighteenth Session of the Legislature of the State of Nevada, 1897. Carson City: State Printing Office, 1897.

The Journal of the Senate of the Eighteenth Session of the Legislature of the State of Nevada, 1897. Carson City: State Printing Office, 1897.

Books

Achard, Steve, and Conrad Buedel. *Lost Legacy of Carson Valley: The Rise and Fall of the H. F. Dangberg Ranching Empire.* Minden, Nev.: Dangberg Partners, 2011.

Adams, Romanzo. *Taxation in Nevada: A History.* Reno: Nevada Historical Society, 1918.

Angulo Morales, Alberto, and Óscar Álvarez Gila, coords. *Las migraciones vascas en perspectiva histórica (siglos XVI–XX).* Bilbao: Universidad del País Vasco, 2002.

Austin, Mary. *The Flock*. Boston: Houghton Mifflin Co., 1906.

Avella, Steven M. *Sacramento and the Catholic Church: Shaping a Capital City*. Reno: University of Nevada Press, 2008.

Azcona Pastor, José Manuel. *Possible Paradises: Basque Emigration to Latin America*. Reno: University of Nevada Press, 2004.

Azkue, Resurrección María de. *Ardi Galdua*. Bilbao: Corazón de Jesús, 1918.

Bancroft, Hubert H. *History of Nevada, Colorado, and Wyoming, 1540–1888*. Vol. 2 of *The Works of Hubert Howe Bancroft*. San Francisco, Calif.: History Company, 1890.

Barnard, Alan, and Jonathan Spencer, eds. *Encyclopedia of Social and Cultural Anthropology*. London and New York: Routledge, 1996.

Bayor, Ronald H. *Encountering Ellis Island: How European Immigrants Entered America*. Baltimore, Md.: Johns Hopkins University Press, 2014.

Bieter, John, and Mark Bieter. *An Enduring Legacy: The Story of Basques in Idaho*. Reno: University of Nevada Press, 2000.

Blakey, Roy G., and Gladys C. Blakey. *The Federal Income Tax*. New York: Longmans, Green, 1940.

Bodnar, John. *Immigration and Industrialization: Ethnicity in an American Mill Town, 1870–1940*. Pittsburgh: University of Pittsburgh Press, 1977.

Bosch, Aurora. *Miedo a la democracia: Estados Unidos ante la Segunda República y la guerra civil española*. Barcelona: Crítica, 2012.

Bukowczyk, John J. *And My Children Did Not Know Me: A History of the Polish-Americans*. Bloomington and Indianapolis: Indiana University Press, 1987.

Bushnell, Eleanor. *The Nevada Constitution: Origin and Growth*. Reno: University of Nevada Press, 1965.

Butler, Anne M. *Daughters of Joy, Sisters of Misery: Prostitutes in the American West, 1865–1890*. Urbana and Chicago: University of Illinois Press, 1985.

Calef, Wesley. *Private Grazing and Public Lands: Studies of the Local Management of the Taylor Act*. Chicago: University of Chicago Press, 1960.

Calvo Salgado, Luís M., María José Fernández Vicente, Axel Kreienbrink, Carlos Sanz Díaz, and Gloria Sanz Lafuente. *Historia del Instituto Español de Emigración: La política migratoria exterior de España y el IEE del Franquismo a la Transición*. Madrid: Editorial Síntesis, 2009.

Cannato, Vincent J. *American Passage: The History of Ellis Island*. New York: Harper Collins, 2009.

Carroll, Peter N., and James D. Fernández, eds. *Facing Fascism: New York and the Spanish Civil War*. New York: Museum of the City of New York and New York University Press, 2007.

Caute, David. *The Great Fear: The Anti-Communist Purge under Truman and Eisenhower*. New York: Simon & Schuster, 1978.

Cava Mesa, María Jesús. *Los diplomáticos de Franco: J. F. de Lequerica, temple y tenacidad (1890–1963)*. Bilbao: Universidad de Deusto, 1989.

Cawley, R. McGreggor. *Federal Land, Western Anger: The Sagebrush Rebellion and Environmental Politics*. Lawrence: University Press of Kansas, 1993.

Cinel, Dino. *From Italy to San Francisco: The Immigrant Experience*. Stanford, Calif.: Stanford University Press, 1982.

Colá y Goiti, José. *La Emigración Vasco-Navarra*. Vitoria: Diputación de Álava, 1882.

Cronon, William. *Nature's Metropolis: Chicago and the Great West*. New York: W.W. Norton, 1991.

Crosby, S. J., Donald F. *God, Church, and Flag: Senator Joseph R. McCarthy and the Catholic Church, 1950–1957*. Chapel Hill: University of North Carolina Press, 1978.

Dallek, Robert. *Franklin D. Roosevelt and American Foreign Policy, 1932–1945*. New York: Oxford University Press, 1979.

Danbom, David B. *The Resisted Revolution: Urban America and the Industrialization of Agriculture, 1900-1930*. Ames: Iowa State University Press, 1979.

Daniels, Roger. *Coming to America: A History of Immigration and Ethnicity in American Life*. New York: Perennial, 2002.

Davies, Richard O., ed. *The Maverick Spirit: Building the New Nevada*. Reno: University of Nevada Press, 1999.

Davis, Sam P. *The History of Nevada:* Vol. I. Reno, Nev.: Elms Publishing Company, 1913.

De la Granja, José Luis, Santiago de Pablo, and Coro Rubio Pobes. *Breve Historia de Euskadi: De los fueros a la autonomía*. Barcelona: Debate, 2011.

Delgado, Lorenzo, and María Dolores Elizalde, eds. *España y Estados Unidos en el siglo XX*. Madrid: CSIC, 2005.

Delgado, Richard, and Jean Stefancic, eds. *Critical White Studies: Looking Behind the Mirror*. Philadelphia: Temple University Press, 1997.

Douglass, William A., and Jon Bilbao. *Amerikanuak: Basques in the New World*. Reno: University of Nevada Press, 1975.

Douglass, William A., Richard W. Etulain, and William H. Jacobsen, Jr., eds. *Anglo-American Contributions to Basque Studies: Essays in Honor of Jon Bilbao*. Reno: Desert Research Institute Publications on the Social Sciences, 1977.

Drago, Harry Sinclair. *Following the Grass*. New York: The Macaulay Company, 1924.

Dykstra, Robert R. *The Cattle Towns*. New York: Alfred A. Knopf, 1968.

Echeverria, Jeronima. *Home Away from Home: A History of Basque Boardinghouses*. Reno: University of Nevada Press, 1999.

Edwards, Jerome E. *Pat McCarran: Political Boss of Nevada*. Reno: University of Nevada Press, 1982.

Edwards, Jill. *Anglo-American Relations and the Franco Question, 1945–1955*. Oxford: Clarendon Press, 1999.

Elliott, Gary E. *Senator Alan Bible and the Politics of the New West*. Reno: University of Nevada Press, 1994.

Elliott, Russell R. *Nevada's Twentieth-Century Mining Boom: Tonopah, Goldfield, Ely*. Reno: University of Nevada Press, 1966.

Elliott, Russell R., with the assistance of William D. Rowley. *History of Nevada*. Lincoln: University of Nebraska Press, 1973.

Embry, Jessie L., and Brian Q. Cannon, eds. *Immigrants in the Far West: Historical Identities and Experiences*. Salt Lake City: University of Utah Press, 2015.

Emmons, David M. *The Butte Irish: Class and Ethnicity in an American Mining Town, 1875–1925*. Urbana: University of Illinois Press, 1989.

Etulain, Richard W., ed. *Basques of the Pacific Northwest*. Pocatello: Idaho State University Press, 1991.

Etulain, Richard W., and Jeronima Echeverria, eds. *Portraits of Basques in the New World*. Reno: University of Nevada Press, 1999.

Eugenia Lida, Clara, comp. *Una inmigración privilegiada: comerciantes, empresarios y profesionales españoles en México en los siglos XIX y XX*. Madrid: Alianza, 1994.

Ewen, Elizabeth. *Immigrant Women in the Land of Dollars*. New York: Monthly Review Press, 1985.

Ferguson, Niall. *Colossus: The Price of America's Empire*. New York: Penguin Press, 2004.

Fiero, Bill. *Geology of the Great Basin*. Reno: University of Nevada Press, 1986.

Fitzgerald, Keith. *The Face of the Nation: Immigration, the State, and the National Identity*. Stanford, Calif.: Stanford University Press, 1996.

Foner, Eric. *Free Soil, Free Labor, Free Men: The Ideology of the Republican Party before the Civil War*. New York: Oxford University Press, 1970.

Foss, Phillip O. *Politics and Grass*. Seattle: University of Washington Press, 1960.

Fusi Aizpurúa, Juan Pablo. *España, 1808–1996: El desafío de la modernidad*. Madrid: Espasa, 1997.

Gallego, Ferran. *El evangelio fascista: La formación de la cultura política del franquismo (1930–1950)*. Barcelona: Crítica, 2014.

Gallop, Rodney. *A Book of the Basques*. London: Macmillan, 1930.

Garate, Donald T. *Juan Bautista de Anza: Basque Explorer in the New World, 1693–1740*. Reno: University of Nevada Press, 2005.

Garland, Hamlin. *Cavanagh, Forest Ranger: A Romance of the Mountain West*. New York: Harper and Brothers, 1910.

Georgetta, Clel. *Golden Fleece in Nevada*. Reno, Nev.: Venture Publishing Co., 1972.

Giralt C., J., ed. *Los Vascos en América. Tomo I: Chile, Argentina y Uruguay*. Santiago de Chile: Cathalonia, 1937.

Glas, Eduardo Jorge. *Bilbao's Modern Business Elite*. Reno: University of Nevada Press, 1997.

González Portilla, Manuel. *Estado, Capitalismo y desequilibrios regionales (1845–1900). Andalucía, País Vasco*. San Sebastián: Haranburu/Universidad del País Vasco, 1985.

González Portilla, Manuel, Manuel Montero García, José María Garmendia Urdangarín, Pedro A. Novo López, and Olga Macías. *Ferrocarriles y desarrollo. Red y mercados en el País Vasco (1856–1914)*. Bilbao: Servicio Editorial de la Universidad del País Vasco, 1996.

Goodman, David. *Spanish Naval Power, 1589–1665: Reconstruction and Defeat*. Cambridge and New York: Cambridge University Press, 1997.

Grover, David H. *Diamondfield Jack: A Study in Frontier Justice*. Norman: University of Oklahoma Press, 1986.

Guirao, Fernando. *Spain and the Reconstruction of Western Europe, 1945–1957: Challenge and Response*. London: Macmillan, 1998.

Hall, Shawn. *Old Heart of Nevada: Ghost Towns and Mining Camps of Elko County*. Reno: University of Nevada Press, 1998.

Handlin, Oscar. *The Uprooted: The Epic Story of the Great Migrations that Made the American People*. Boston, Mass.: Little, Brown and Co., 1951.

Hays, Samuel P. *Conservation and the Gospel of Efficiency: The Progressive Conservation Movement, 1890–1920*. Cambridge, Mass.: Harvard University Press, 1959.

Heyer, Kristen E., Mark J. Rozell, and Michael A. Genovese, eds. *Catholics and Politics: The Dynamic Tension between Faith and Power*. Washington, D.C.: Georgetown University Press, 2008.

Higham, John. *Strangers in the Land: Patterns of American Nativism, 1860–1925*. New Brunswick, NJ: Rutgers University Press, 1955.

Hofstadter, Richard. *The Progressive Historians: Turner, Beard, Parrington*. New York: Alfred A. Knopf, 1968.

Houghton, Samuel G. *A Trace of Desert Waters: The Great Basin Story*. Salt Lake City, Utah: Howe Brothers, 1986.

Hulse, James W. *The Silver State: Nevada's Heritage Reinterpreted*. Reno: University of Nevada Press, 1991.

Igler, David. *Industrial Cowboys: Miller & Lux and the Transformation of the Far West, 1850–1920*. Berkeley: University of California Press, 2001.

Irigaray, Louis, and Theodore Taylor. *A Shepherd Watches, A Shepherd Sings*. Garden City, New York: Doubleday and Co., 1977.

Irujo, Xabier. *Expelled from the Motherland: The Government of President Jose Antonio Agirre in Exile (1937–1960)*. Reno, Nev.: Center for Basque Studies Press, 2012.

Israel, Fred L. *Nevada's Key Pittman*. Lincoln: University of Nebraska Press, 1963.

Jacobson, Matthew F. *Special Sorrows: The Diasporic Imagination of Irish, Polish, and Jewish Immigrants in the United States*. Cambridge, Mass.: Harvard University Press, 1995.

———. *Whiteness of a Different Color: European Immigrants and the Alchemy of Race*. Cambridge, Mass.: Harvard University Press, 1998.

Jacoby, Karl. *Crimes against Nature: Squatters, Poachers, Thieves, and the Hidden History of American Conservation*. Berkeley: University of California Press, 2001.

James, Ronald M. *The Roar and the Silence: A History of Virginia City and the Comstock Lode*. Reno: University of Nevada Press, 1998.

James, Ronald M., and C. Elizabeth Raymond, eds. *Comstock Women: The Making of a Mining Community*. Reno: University of Nevada, Reno, 1998.

Jameson, Elizabeth, and Susan Armitage, eds. *Writing the Range: Race, Class, and Culture in the Women's West*. Norman: University of Oklahoma Press, 1997.

Johnson, Chalmers. *The Sorrows of Empire: Militarism, Secrecy, and the End of the Republic*. New York: Metropolitan Books, 2004.

Jordan, Terry G., and Matti Kaups. *The American Backwoods Frontier: An Ethnic and Ecological Interpretation*. Baltimore, Md.: Johns Hopkins University Press, 1989.

Juliá Díaz, Santos, ed. *Socialismo y guerra civil*. Madrid: Fundación Pablo Iglesias, 1987.

Kennedy, David M. *Freedom from Fear: The American People in Depression and War, 1929–1945*. New York: Oxford University Press, 1999.

Kessell, John L. *Spain in the Southwest: A Narrative History of Colonial New Mexico, Arizona, Texas, and California*. Norman: University of Oklahoma Press, 2002.

King, Desmond S. *Making Americans: Immigration, Race, and the Origins of Diverse Democracy*. Cambridge, Mass.: Harvard University Press, 2000.

Kroes, Rob. *The Persistence of Ethnicity: Dutch Calvinist Pioneers in Amsterdam, Montana*. Urbana: University of Illinois Press, 1992.

Langston, Nancy. *Where Land and Water Meet: A Western Landscape Transformed*. Seattle and London: University of Washington Press, 2003.

Laxalt, Paul. *Nevada's Paul Laxalt: A Memoir*. Reno, Nev.: Jack Bacon & Co, 2000.

León Aguinaga, Pablo. *Sospechosos habituales, El cine norteamericano, Estados Unidos y la España franquista, 1939–1960*. Madrid: Consejo Superior de Investigaciones Científicas, 2010.

Lhande, Pierre. *L'Emigration Basque: Histoire, économie, psychologie*. Paris: Nouvelle Librairie Nationale, 1910.

Liedtke, Boris N. *Embracing a Dictatorship: US Relations with Spain, 1945–53*. New York: St. Martin's Press, 1998.

Limerick, Patricia N. *The Legacy of Conquest: The Unbroken Past of the American West*. New York: W.W. Norton, 1987.

Luebke, Frederick C., ed. *European Immigrants in the American West: Community Histories*. Albuquerque: University of New Mexico Press, 1998.

———, ed. *Germans in the New World: Essays in the History of Immigration*. Chicago: University of Illinois Press, 1990.

———. *Immigrants and Politics: The Germans of Nebraska, 1880–1900*. Lincoln: University of Nebraska Press, 1969.

Mack, Effie M. *The Indian Massacre of 1911 at Little High Rock Canyon, Nevada*. Sparks, Nev.: Western Print and Publishing Company, 1968.

———. *Nevada: A History of the State from the Earliest Times through the Civil War*. Glendale, Calif.: The Arthur H. Clark Company, 1936.

Mallea-Olaetxe, Jose. *Speaking through the Aspens: Basque Tree Carvings in California and Nevada*. Reno: University of Nevada Press, 2000.

McLaughlin, Virginia Y. *Family and Community: Italian Immigrants in Buffalo, 1880–1930*. Urbana and Chicago: University of Illinois Press, 1977.

McPhee, John. *Basin and Range*. New York: Farrar, Strauss, and Giroux, 1981.

McWilliams, Carey. *Factories in the Field: The Story of Migratory Farm Labor in California*. Boston: Little, Brown and Co., 1939.

Merrill, Karen R. *Public Lands and Political Meaning: Ranchers, the Government, and the Property between Them*. Berkeley: University of California Press, 2002.

Meyerson, Harvey. *Nature's Army: When Soldiers Fought for Yosemite*. Lawrence: University Press of Kansas, 2001.

Miller, Char. *Gifford Pinchot and the Making of Modern Environmentalism*. Washington, D.C.: Island Press, 2001.

Miller, Douglas T., and Marion Nowak. *The Fifties: The Way We Really Were*. Garden City, New York: Doubleday and Company, 1977.

Miller, Mary Catherine. *Flooding the Courtrooms: Law and Water in the Far West*. Lincoln: University of Nebraska Press, 1993.

Moore, Jacqueline M. *Cow Boys and Cattle Men: Class and Masculinities on the Texas Frontier, 1865–1900*. New York: New York University Press, 2010.

Moradiellos, Enrique. *La España de Franco (1939–1975): Política y sociedad*. Madrid: Editorial Síntesis, 2003.

Morawska, Ewa T. *For Bread with Butter: Life-Worlds of East Central Europeans in Johnstown, Pennsylvania, 1890–1940*. New York: Cambridge University Press, 1985.

Moskos, Charles C. Jr., *Greek Americans: Struggle and Success*. Englewood Cliffs, N.J.: Prentice Hall, 1980.

Mota Zurdo, David. *Un sueño americano: El Gobierno Vasco en el exilio y Estados Unidos (1937–1979)*. Oñati: IVAP, 2016.

Muir, John. *The Mountains of California*. New York: The Century Co., 1907.

Murphy, J. Carter, and R. Richard Rubottom. *Spain and the United States since World War II*. New York: Praeger, 1984.

Nash, Gerald D. *The American West Transformed: The Impact of the Second World War*. Lincoln: University of Nebraska Press, 1985.

Ngai, Mae M. *Impossible Subjects: Illegal Aliens and the Making of Modern America*. Princeton, N.J.: Princeton University Press, 2004.

Nicolás Marín, María Encarna, and María Alicia Alted Vigil. *Disidencias en el franquismo (1939–1975)*. Murcia: DM, 1999.

Niño Rodríguez, Antonio, and José Antonio Montero Jiménez, eds. *Guerra fría y propaganda. Estados Unidos y su cruzada cultural en Europa y América Latina*. Madrid: Biblioteca Nueva, 2012.

Olzak, Susan, and Joane Nagel, ed. *Competitive Ethnic Relations*. Orlando, FL: Academic Press, 1986.

O'Neal, Bill. *Cattlemen vs. Sheepherders: Five Decades of Violence in the West, 1880–1920*. Austin, Tex.: Eakin Press, 1989.

Oshinsky, David M. *A Conspiracy So Immense: The World of Joe McCarthy*. New York: Free Press, 1983.

Ostergren. Robert C. *A Community Transplanted: The Trans-Atlantic Experience of a Swedish Immigrant Settlement in the Upper Middle West, 1835–1915*. Madison: University of Wisconsin Press, 1988.

Ostrander, Gilman M. *Nevada: The Great Rotten Borough, 1859–1964*. New York: Alfred A. Knopf, 1966.

Paris, Beltran, and William A. Douglass. *Beltran: Basque Sheepman of the American West*. Reno: University of Nevada Press, 1979.

Patterson, Edna B., Louise A. Ulph, and Victor Goodwin. *Nevada's Northeast Frontier*. Sparks, Nev.: Western Printing and Publishing Company, 1969.

Payne, Stanley G. *Falange. The Franco Regime, 1936–1975*. Madison: University of Wisconsin Press, 1987.

———. *A History of Spanish Fascism*. Stanford, Calif.: Stanford University Press, 1961.

Peck, Gunther. *Reinventing Free Labor: Padrones and Immigrant Workers in the North American West, 1880–1930*. New York: Cambridge University Press, 2000.

Peffer, Louise E. *The Closing of the Public Domain: Disposal and Reservation Policies, 1900–1950*. Stanford, Calif.: Stanford University Press, 1951.

Pildain Salazar, María Pilar. *Ir a América: La emigración vasca a América, Guipúzcoa, 1840–1870*. San Sebastián: Sociedad Guipuzcoana de Ediciones y Publicaciones, 1984.

Pisani, Donald J. *To Reclaim a Divided West: Water, Law, and Public Policy, 1848–1902*. Albuquerque: University of New Mexico Press, 1992.

Portero, Florentino. *Franco aislado: La cuestión española, 1945–1950*. Madrid: Aguilar, 1989.

Powaski, Ronald E. *The Cold War: The United States and the Soviet Union, 1917–1991*. New York: Oxford University Press, 1998.

Powers, Richard Gid. *Not Without Honor: The History of American Anticommunism*. New York: Free Press, 1995.

Powers, Stephen. *The American Merino: For Wool and for Mutton*. New York: Orange Judd Company, 1907.

Ratner, Sidney, James H. Soltow, and Richard Sylla. *The Evolution of the American Economy: Growth, Welfare, and Decision Making*. New York: Basic Books, 1979.

Ripley, William Z. *The Races of Europe: A Sociological Study*. New York: D. Appleton and Company, 1899.

Roberts, Peter. *The New Immigration: A Study of the Industrial and Social Life of Southeastern Europeans in America*. New York: Macmillan, 1913.

Robbins, William G., and James C. Foster, eds. *Land in the American West: Private Claims and the Common Good*. Seattle and London: University of Washington Press, 2000.

Roediger, David R. *The Wages of Whiteness: Race and the Making of the American Working Class*. London and New York: Verso, 1991.

Roosevelt, Theodore. *The Winning of the West*, Vol. 1. New York: G.P. Putman's Sons, 1889.

Rosswurm, Steve. *The FBI and the Catholic Church, 1935–1962*. Amherst: University of Massachusetts Press, 2009.

Rowe, Vivian. *The Basque Country*. London: Putman, 1955.

Rowley, William D. *Reclaiming the Arid West: The Career of Francis G. Newlands*. Bloomington: Indiana University Press, 1996.

———. *U.S. Forest Service Grazing and Rangelands: A History*. College Station: Texas A&M University Press, 1985.

Rubio Pobes, Coro. *Fueros y constitución: La lucha por el control del poder, País Vasco, 1808–1868*. Bilbao: Universidad del País Vasco, 1997.

———. *La identidad vasca en el siglo XIX: Discurso y agentes sociales*. Madrid: Biblioteca Nueva, 2003.

———. *Revolución y tradición: El País Vasco ante la Revolución liberal y la construcción del Estado español, 1808–1868.* Madrid: Siglo XXI, 1996.

San Sebastián, Koldo. *El exilio vasco en América: 1936/1946–Acción del Gobierno.* San Sebastián: Txertoa, 1988.

Sawyer, Byrd W. *Nevada Nomads: A Story of the Sheep Industry.* San Jose, Calif.: Harlan-Young Press, 1971.

Schlebecker, John T. *Whereby We Thrive: A History of American Farming, 1607–1972.* Ames: Iowa State University Press, 1975.

Schrecker, Ellen. *Many Are the Crimes: McCarthyism in America.* Boston, Mass.: Little, Brown, and Company, 1998.

Shanks, Cheryl. *Immigration and Politics of American Sovereignty, 1890–1990.* Ann Arbor: University of Michigan Press, 2001.

Shepherd, William. *Prairie Experiences in Handling Cattle and Sheep.* London: Chapman and Hall, 1884.

Shepperson, Wilbur S. *Restless Strangers: Nevada's Immigrants and Their Interpreters.* Reno: University of Nevada Press, 1970.

Silen, Sol. *La historia de los Vascongados en el oeste de los Estados Unidos.* New York: Las Novedades, 1917.

Simmons, Marc. *The Last Conquistador: Juan de Oñate and the Settling of the Far Southwest.* Norman: University of Oklahoma Press, 1991.

Small, Stephen. *Racialised Barriers: The Black Experience in the USA and England in the 1980s.* London and New York: Routledge, 1994.

Smith, Henry N. *Virgin Land: The American West as Symbol and Myth.* Cambridge, Mass.: Harvard University Press, 1950.

Smith, Moroni A. *Herding and Handling Sheep: On the Open Range in U.S.A.* Salt Lake City, Utah: n.p., 1918.

Steen, Harold K, ed. *Forest and Wildlife Science in America: A History.* Durham, N.C.: Forest History Society, 1999.

———, ed. *History of Sustained-Yield Forestry: A Symposium.* Durham, N.C.: Forest History Society, 1984.

———. *The U.S. Forest Service: A History.* Seattle: University of Washington Press, 1976.

Stein, Harold, ed. *American Civil-Military Decisions: A Book of Case Studies.* Birmingham: University of Alabama Press, 1963.

Stewart, Henry. *The Shepherd's Manual: A Practical Treatise on the Sheep; Designed Especially for American Shepherds.* New York: O. Judd Co., 1876.

Szasz, Ferenc M. *Religion in the Modern American West.* Tucson: University of Arizona Press, 2000.

Taussig, Frank W. *The Tariff History of the United States.* New York: Putnam, 1910.

Tchudi, Stephen, ed. *Change in the American West: Exploring the Human Dimension.* Reno: University of Nevada Press, 1996.

———, ed. *Community in the American West.* Reno and Las Vegas: Nevada Humanities Committee, 1999.

Thomàs, Joan María. *La Falange de Franco: Fascismo y fascistización en el régimen franquista (1937–1945)*. Barcelona: Plaza & Janés Editores, 2001.

———. *Roosevelt and Franco during the Second World War: From the Spanish Civil War to Pearl Harbor*. New York: Palgrave Macmillan, 2008.

———. *Roosevelt, Franco, and the End of the Second World War*. New York: Palgrave Macmillan, 2011.

Tierney, Dominic. *FDR and the Spanish Civil War: Neutrality and Commitment in the Struggle That Divided America*. Durham and London: Duke University Press, 2007.

Torres, Rodolfo D., Louis F. Mirón, and Jonathan Xavier Inda, eds. *Race, Identity and Citizenship: A Reader*. Malden, Mass.: Blackwell, 1999.

Totoricagüena, Gloria P. *Identity, Culture, and Politics in the Basque Diaspora*. Reno: University of Nevada Press, 2004.

Towne, Charles W., and Edward N. Wentworth. *Shepherd's Empire*. Norman: University of Oklahoma Press, 1945.

Townley, John M. *Turn This Water into Gold: The Story of the Newlands Project*. Reno: Nevada Historical Society, 1998.

Truett, Velma Stevens. *On the Hoof in Nevada*. Los Angeles, Calif.: Lorrin L. Morrison, 1950.

Turner, Frederick Jackson. *The Frontier in American History*. New York: Henry Holt and Company, 1948.

Tusell, Javier. *Franco y los católicos: La política interior española entre 1945 y 1957*. Madrid: Alianza Editorial, 1984.

Tusell, Javier, Juan Avilés, and Rosa Pardo, eds. *La Política Exterior de España en el Siglo XX*. Madrid: Biblioteca Nueva, 2000.

Ugalde, Alexander, coord. *Patria y libertad: Los vascos y las guerras de independencia de Cuba (1868–1898)*. Tafalla: Txalaparta, 2012.

Utley, Robert M. *The Indian Frontier of the American West, 1846–1890*. Albuquerque: University of New Mexico Press, 1984.

Vaught, David. *Cultivating California: Growers, Specialty Crops, and Labor, 1875–1920*. Baltimore, Md.: Johns Hopkins University Press, 1999.

Viñas, Ángel. *En las garras del águila: Los pactos con Estados Unidos, de Francisco Franco a Felipe González (1945–1995)*. Barcelona: Crítica, 2003.

———. *Guerra, dinero, dictadura: Ayuda fascista y autarquía en la España de Franco*. Barcelona: Crítica, 1984.

———. *Los pactos secretos de Franco con Estados Unidos: Bases, ayuda económica y recortes de soberanía*. Barcelona: Grijalbo, 1981.

Wagnleitner, Reinghold, and Elaine Tyler May, eds. *"Here, There and Everywhere": The Foreign Policy of American Popular Culture*. Hanover, N.H.: University Press of New England, 2000.

Walton, Clifford C. *Capitol's Who's Who for Nevada, 1949–1950*. Portland, Ore.: Capitol Publishing Company, 1949.

Webb, Walter Prescott. *The Great Plains*. Lincoln: University of Nebraska Press, 1981.

Weber, David J. *The Spanish Frontier in North America*. New Haven, Conn.: Yale University Press, 1992.

Welles, Summer. *The Time for Decision*. New York: Harper & Brothers, 1944.

Wentworth, Edward N. *America's Sheep Trails: History, Personalities*. Ames: Iowa State College Press, 1948.

White, Richard. *"It's Your Misfortune and None of My Own": A History of the American West*. Norman: University of Oklahoma Press, 1991.

Wiebe, Robert H. *The Search for Order, 1877–1920*. New York: Hill and Wang, 1967.

Wittke, Carl. *We Who Built America: The Saga of the Immigrant*. New York: Prentice-Hall, 1940.

Woolen, David B., and Richard G. Kurial, eds. *FDR, the Vatican, and the Roman Catholic Church in America, 1933–1945*. New York: Palgrave Macmillan, 2003.

Wrobel, David M., and Michael C. Steiner, eds. *Many Wests: Place, Culture, and Regional Identity*. Lawrence: University Press of Kansas, 1997.

Ybarra, Michael J. *Washington Gone Crazy: Senator Pat McCarran and the Great American Communist Hunt*. Hanover, N.H.: Steerforth Press, 2004.

Young, James A., and B. Abbott Sparks. *Cattle in the Cold Desert*. Logan: Utah State University Press, 1985.

Zanjani, Sally. *Goldfield: The Last Gold Rush on the Western Frontier*. Athens: Swallow Press/Ohio University Press, 1992.

Zolberg, Aristide R. *A Nation by Design: Immigration Policy in the Fashioning of America*. Cambridge, Mass.: Harvard University Press with Russell Sage Foundation, 2006.

Zulaika, Joseba. *Del Cromañón al carnaval: Los vascos como museo antropológico*. San Sebastián: Erein, 1996.

Articles

Adams, Romanzo. "Public Range Lands: A New Policy Needed." *American Journal of Sociology* 22, no. 3 (November 1916): 324–51.

Azcona Mauleon, Jesús. "La delimitación antropológica y etnológica de lo vasco y de los vascos." *Cuadernos de etnología y etnografía de Navarra* 14, no. 40 (1982): 753–802.

"Basque Shepherd: U.S. Immigrant Works Amid the Alien Lambs." *Life* 32, no. 19 (May 12, 1952): 124–29.

Bieter, Patrick. "Reluctant Shepherds: The Basques in Idaho." *Idaho Yesterdays* 1 (Summer 1957): 10–15.

Bosch, Aurora. "Entre la democracia y la neutralidad: Estados Unidos ante la Guerra Civil Española." *Ayer* 90, no. 2 (2013): 167–87.

———. "Estados Unidos en los años treinta: ¿Un socialismo imposible?" *Historia Social*, no. 11 (Fall 1991): 39–56.

Boyd, Monica. "Family and Personal Networks in International Migration: Recent Developments and New Agendas." *International Migration Review* 23, no. 3 (Fall 1989): 638–70.

Bradford, Saxton E. "Sons of the Pyrenees in the Northwest." *Travel* 78 (September 1942): 11–15.

Brodhead, Michael J. "Accepting the Verdict: National Supremacy as Expressed in State Constitutions, 1861–1912." *Nevada Historical Society Quarterly* 13, no. 2 (Summer 1970): 3–16.

Brodhead, Michael J., and James W. Hulse. "Paul W. Gates, Western Land Policy and the Equal Footing Doctrine." *Nevada Historical Society Quarterly* 29, no. 4 (Winter 1986): 225–40.

Calderon, Rosario and Esther Rebato. "Historia de la antropología biológica en el País Vasco." *Revista Internacional de los Estudios Vascos* 42, no. 1 (1997): 45–62.

Calvo González, Oscar. "¡Bienvenido, Míster Marshall! La ayuda económica americana y la economía española en la década de 1950." *Revista de Historia Económica* 19 (2001): 253–76.

Campbell, Craig. "The Basque-American Ethnic Area: Geographical Perspectives on Migration, Population, and Settlement." *Journal of Basque Studies* 6 (1985): 83–89.

Collado Seidel, Carlos. "¿De Hendaya a San Francisco? Londres y Washington contra Franco y la Falange (1942–1945)." *Espacio, Tiempo y Forma, Serie V, Historia Contemporánea* no. 7 (1994): 51–84.

Creel, George. "Under the Underdog." *Collier's* 95, no. 14 (April 6, 1935): 26.

Darling, Arthur B. "Grover Cleveland and National Finances." *American Scholar* 3, no. 2 (Spring 1934): 144–54.

"Death on the Range." *Time* 53, no. 6 (February 7, 1949): 13–14.

Delgado Gómez-Escalonilla, Lorenzo. "¿El 'amigo americano'? España y Estados Unidos durante el franquismo." *Studia Histórica* 21 (2003): 231–76.

Douglass, William A. "The Basques of the American West: Preliminary Historical Perspectives." *Nevada Historical Society Quarterly* 13, no. 4 (Winter 1970): 12–25.

———. "Rural Exodus in Two Spanish Basque Villages: A Cultural Explanation." *American Anthropologist* 73, no. 5 (October 1971): 1100–14.

———. "Lonely Lives under the Big Sky." *Natural History* 82, no. 3 (March 1973): 28–39.

Echeverria, Jeronima. "Basque 'Tramp Herders' on Forbidden Ground: Early Grazing Controversies in California's National Reserves." *Locus* 4, no. 1 (Fall 1991): 41–58.

Edwards, Jerome E. "Nevada Power Broker: Pat McCarran and His Political Machine." *Nevada Historical Society Quarterly* 27, no. 3 (Fall 1984): 182–98.

Fanjul, Enrique. "El papel de la ayuda americana en la economía española, 1951–1957." *Información Comercial Española* no. 577 (September 1981): 159–65.

Fox Maura, Soledad. "Miradas Opuestas: La Casa Blanca y la opinión pública norteamericana ante la guerra de España." *Circumstancia* 19 (Mayo 2009).

Garate, Donald T. "Basque Names, Nobility, and Ethnicity on the Spanish Frontier." *Colonial Latin American Historical Review* 2, no. 1 (Winter 1993): 77–104.

———. "Juan de Oñate's *Prueba de Caballero*, 1625: A Look at His Ancestral Heritage." *Colonial Latin American Historical Review* 7, no. 2 (Spring 1998): 129–73.

Georgetta, Clel. "Sheep in Nevada." *Nevada Historical Society Quarterly* 8, no. 2 (Summer 1965): 17–38.

Griffith, Ernest S. "The Place of Congress in Foreign Relations." *Annals of the American Academy of Political and Social Science* 289 (September 1953): 11–21.

Handlin, Oscar. "The Immigration Fight Has Just Begun." *Commentary* 14 (July 1952): 1–7.

Harkness, Ione B. "Basque Settlement in Oregon." *Oregon Historical Quarterly* 34, no. 3 (September 1933): 273–75.

Hatfield, Kevin D. "'We Were Not Tramp Sheepmen': Joe Odiaga and Acculturation, Resistance, and Identity in the Bizkaian Basque Community, 1890–1946." *Nevada Historical Society Quarterly* 52, no. 4 (Winter 2009): 292–312.

Hernández Sandoica, Elena. "La Compañía Trasatlántica Española: Una dimensión ultramarina del capitalismo español." *Historia contemporánea* 2 (1989): 119–64.

Howard, Randall R. "The Sheep-Herder." *Outlook* 96 (September–December 1910): 943–50.

Igler, David. "Industrial Cowboys: Corporate Ranching in Late Nineteenth-Century California." *Agricultural History* 69, no. 2 (Spring 1995): 201–15.

Jordan, Terry G. "New Sweden's Role on the American Frontier: A Study in Cultural Preadaptation." *Geografiska Annaler* 71, no. 2 (1989): 71–83.

———. "Preadaptation and European Colonization in Rural North America." *Annals of the Association of American Geographers* 79, no. 4 (December 1989): 489–500.

Lane, Richard. "Basque Tree Carvings." *Northeastern Nevada Historical Society Quarterly* 1, no. 3 (Winter 1971): 1–9.

Light, Ivan. "Immigrant and Ethnic Enterprise in North America." *Ethnic and Racial Studies* 7, no. 2 (April 1984): 195–216.

McCarran, Margaret P. "Patrick Anthony McCarran, 1876–1954," (part 1). *Nevada Historical Society Quarterly* 11, no. 3–4 (Fall/Winter 1968): 5–66.

———. "Patrick Anthony McCarran, 1876–1954," (part 2). *Nevada Historical Society Quarterly* 12, no. 1 (Spring 1969): 5–75.

Newton, Milton. "Cultural Preadaptation and the Upland South." *Geoscience and Man* 5 (June 1974): 143–54.

Nicolás Marín, María Encarna. "Notas: La emigración española a América Latina." *Areas,* no. 7 (1986): 99–105.

Olson, James S. "Pioneer Catholicism in Eastern and Southern Nevada, 1864–1931." *Nevada Historical Society Quarterly* 26, no. 3 (Fall 1983): 159–71.

Olla Brundu, Paola. "Europa y Estados Unidos frente al problema de la integración de España en el sistema de seguridad occidental (1945–1953)." *Spagna contemporanea* no. 15 (1999): 93–112.

Orth, Samuel P. "The Alien Contract Labor Law." *Political Science Quarterly* 22, no. 1 (March 1907): 49–60.

Patterson, Edna B. "Early Cattle in Elko County." *Nevada Historical Society Quarterly* 8, no. 2 (Summer 1965): 5–12.

———. "Spanish Ranch Letters." *Northeastern Nevada Historical Society Quarterly* 84, no. 1 (Winter 1984): 3–15.

Perry, Frank V. "The Last Indian Uprising in the United States." *Nevada Historical Society Quarterly* 15, no. 4 (Winter 1972): 23–37.
Pisani, Donald J. "Federal Reclamation and Water Rights in Nevada." *Agricultural History* 51, no. 3 (July 1977): 540–58.
Rakestraw, Lawrence. "Sheep Grazing in the Cascade Range: John Minto vs. John Muir." *Pacific Historical Review* 27, no. 4 (November 1958): 371–82.
Reder, Melvin W. "The Theory of Occupational Wage Differentials." *American Economic Review* 45, no. 5 (December 1955): 833–852.
Rhoades, Robert E. "Foreign Labor and German Industrial Capitalism, 1871–1978: The Evolution of a Migratory System." *American Ethnologist* 5, no. 3 (August 1978): 553–73.
Rubio Pobes, Coro. "Centinelas de la Patria: Regionalismo vasco y nacionalización española en el siglo XIX." *Historia contemporánea*, no. 53 (2016): 393–425.
Saitua, Iker. "Becoming Herders: Basque Immigration, Labor, and Settlement in Nevada, 1880–1910." *Montana, The Magazine of Western History* 66, no. 4 (Winter 2016): 58–70.
———. "*The Best Sheepherder*: The Racial Stereotype of Basque Immigrants in the American West Between the End of the Nineteenth and the Beginning of the Twentieth Centuries." *Historia Contemporánea* 1, no. 56 (2018), 81–119.
———. "*Distilling Spirits*: Inmigrantes vascos, cultura de la bebida y prohibición en el estado de Nevada, 1910–1920." *Historia Social* 1, no. 90 (2018): 45–65.
———. "Resistance to the 'Hoofed Locusts': Key Pittman, Basque Immigrants, and the Sheep Industry of Nevada, 1910–1920." *Nevada Historical Society Quarterly* 60, no. 1–4 (2017): 148–72.
Shinn, Charles H. "Work in a National Forest." *Forestry and Irrigation* 13 (November 1907): 590–97.
Spillman, W. J. "The Agricultural Ladder." *American Economic Review* 9, no. 1 [Suppl.] (March 1919): 170–79.
Steinberg, Alfred. "McCarran: Lone Wolf of the Senate." *Harper's Magazine* (November 1950): 89–95.
Stewart, Halina M. "The Nevada State Detectives." *Nevada Historical Society Quarterly* 49, no. 1 (Spring 2006): 23–41.
Taussig, Frank W. "The Tariff Act of 1897." *Quarterly Journal of Economics* 12, no. 1 (October 1897): 42–69.
———. "The United States Tariff Act of 1897." *The Economic Journal* 7, no. 28 (December 1897): 592–98.
Townley, John M. "Management of Nevada's State Lands, 1864–1900." *Journal of the West* 17, no. 1 (January 1978): 63–73.
Turner, Frederick Jackson. Review of *The Winning of the West*. By Theodore Roosevelt. *The Dial* 10 (August 1889): 71–73.
Wentworth, Edward Norris. "Eastward Sheep Drives from California and Oregon." *Mississippi Valley Historical Review* 28, no. 4 (March 1942): 507–38.
"West Fights Worst Winter in History." *Life* 26, no. 7 (February 14, 1949): 19–23.
Yancey, William L., Eugene P. Ericksen, and Richard N. Juliani. "Emergent Ethnicity:

A Review and Reformulation." *American Sociological Review* 41, no. 3 (June 1976): 391–403.

Young, James A. "Hay Making: The Mechanical Revolution on the Western Range." *Western Historical Quarterly* 14, no. 3 (July 1983): 311–26.

Zellick, Anna. "'We All Intermingled': The Childhood Memories of South Slavic Immigrants in Red Lodge and Bearcreek, Montana, 1904–1943." *Montana, The Magazine of Western History* 44, no. 3 (Summer 1994): 34–45.

Newspapers and Trade Journals

ABC (Madrid, Spain)
Boston Post (Boston, Massachusetts)
California Cultivator and Livestock and Dairy Journal (Los Angeles, California)
California-ko Eskual Herria [California's Basque Country] (Los Angeles, California)
California Livestock News (San Francisco, California)
Daily Free Press (Elko, Nevada)
Daily Ledger (Canton, Illinois)
Daily Nevada State Journal (Reno, Nevada)
Des Moines Daily News (Des Moines, Iowa)
Elko Daily Free Press (Elko, Nevada)
Elko Independent (Elko, Nevada)
Eskualduna [Basque] (Bayonne, France)
Humboldt Star (Winnemucca, Nevada)
Idaho State Journal (Pocatello, Idaho)
Idaho Statesman (Boise, Idaho)
Idaho Wool Growers Bulletin (Boise, Idaho)
Morning Appeal (Carson City, Nevada)
National Wool Grower (Salt Lake City, Utah)
Nevada State Journal (Reno, Nevada)
News (Golconda, Nevada)
New York Times (New York City, New York)
Oregonian (Portland, Oregon)
Reno Evening Gazette (Reno, Nevada)
Salt Lake Daily Tribune (Salt Lake City, Utah)
Salt Lake Tribune (Salt Lake City, Utah)
San Francisco Call (San Francisco, California)
Silver State (Winnemucca, Nevada)
Sunday Star (Washington, D.C.)
Territorial Enterprise (Virginia City, Nevada)
Times-Review (Tuscarora, Nevada)
Walla Walla Bulletin (Walla Walla, Washington)
Washington Post (Washington, D.C.)
The Weekly Gazette and Stockman (Reno, Nevada)

Index

About the Author

Iker Saitua is currently a postdoctoral fellow in history at the University of California, Riverside, and the University of the Basque Country (UPV/EHU). Saitua received his PhD from the University of Nevada, Reno, in 2016. Thereafter, he presented his dissertation at the UPV/EHU, getting a second PhD that received summa cum laude honors in history. A native of the Basque Country, Saitua obtained his BA in history from UPV/EHU in 2010 before earning an MA degree in history, an MA degree in secondary education, and his two PhD degrees. Iker Saitua is a historian of the Basque Country and the United States specializing in the American West, the history of capitalism and commodities, environmental history, and migration and labor history. His work has focused largely on the intersections between agricultural labor, natural resources, and livestock raising in the American West. Saitua is the author of various articles on the Basque immigrant sheepherders in Nevada and the West.